ADVANCED LEARNER'S DICTIONARY *Of* STATISTICS

ADVANCED LEARNER'S DICTIONARY

Of

STATISTICS

Compiled & Edited by

A Team of Experts

ANMOL PUBLICATIONS PVT. LTD.

NEW DELHI - 110 002 (INDIA)

ANMOL PUBLICATIONS PVT. LTD.
4374/4B, Ansari Road, Daryaganj
New Delhi - 110 002

Advanced Learner's Dictionary of Statistics

First Edition, 2000

ISBN 81-261-0488-0

[Responsibility for the facts stated, opinions expressed, conclusions reached and plagiarism, if any, in these volumes is entirely that of the Editor. The Publishers bear no responsibility for them whatsoever.]

PRINTED IN INDIA

Published by J.L. Kumar for Anmol Publications Pvt. Ltd., New Delhi and Printed at Mehra Offset Press, Delhi.

Preface

This *Advanced Learner's Dictionary of Statistics* is basically designed for students and teachers as well. It is the result of a careful analysis of the needs of advanced level students and should be an invaluable aid in their day-to-day learning.

As it is extremely difficult to limit the contents of such a dictionary at 'Advanced' level standard, many headwords are included that are above this level and that may be of interest to Advanced Learners. A few headwords that are not strictly scientific are also included because of their common usage in everyday scientific terminology. It has been our aim to provide explanations that are easily understood, whilst maintaining academic standards.

We have tried to include the more important terms used at advanced level, but would welcome any comments regarding serious omissions or indeed about the content of any of the entries.

We would like to thank all the people who have cooperated in producing this dictionary. We are also grateful to many people who have given additional help and advice.

—Editors

A

Abnormality (Abnormalita): Refers to a measure of dissimilarity from the Normal curve, which is somewhat similar to a measure of kurtosis.

Abrupt distribution: A continuous frequency distribution has been considered to be abrupt at a finite terminal point of its range when the frequency and the first derivative of the frequency have not been both zero at that point.

Abscissa: The horizontal or x coordinate in a two-dimensional system of Cartesian coordinates. The vertical or y-coordinate is termed as the *ordinate.*

Absolute deviation: Refers to the difference which is taken without regard to sign, of a variate value from some given value which way be a constant, a 'parent value or a sample value.

Absolute difference: Refers to the difference which is taken without regard to sign, between the values of two variables; and in particular of two random variables.

Absolute error: The absolute error of an observation x has been the absolute deviation of x from its "true' value.

Absolute frequency: Refers to the actual frequency of a variate, as distinct from the relative frequency, namely the ratio of the frequency to the total frequency of all variate values.

Absolute measure: Refers to a measure of variate values which is independent of origin and scale.

Absolute number: Refers to any number as such, simply as a number, without reference to any particular objects whatever except in so far as these objects possess the number property.

Absolute value of a number: The value of a number without regard to positive or negative (plus or minus) sign. The absolute, value of both + 5 and — 5 has' been 5.

Absolute value or modulus, *n*: Refers to the magnitude of a real number or quantity, regardless of its sign. Symbol: straight brackets | |.

| —7 | = 7 | 18.3 | = 18.3

Abso bing state: The term used for a state In a stochastic process which, once reached, cannot get quitted. It is also called an absorbing barrier, but this term is generally used where the process terminates when the barrier gets reached.

Absorption distributions: Refers to the distributions that represent the number of 'individuals' (e.g., particles) that are unable to cross a specified absorbing region.

Acceptable Quality Level (AQL): The term used for the proportion of effective units in a batch which is considered as desirable by the consumer of the batch; the complement of the proportion of defectives which he wishes to tolerate.

Acceptable Reliability Level: Refers to a concept which is analogous to acceptable quality level, in which the characteristic has been expressed in terms of failure rate per unit time, e.g., 1000 hours.

Acceptance control chart: It is a simple form of quality control chart in which lots have been accepted or rejected in accordance to whether plotted values of the mean value of the lot or batch fall inside or outside the specified acceptance control limits.

Acceptance error: Refers to an alternative name for β error.

Acceptance line: In sequential analysis it refers to the graph of' the acceptance number as ordinate against the sample number as abscissa. It is also termed as the acceptance boundary. There has been a corresponding rejection line.

Acceptance number or allowable defects: Refer to the greatest number of defective, in a sample that will permit the batch (from which the sample is drawn) to be accepted without further quality control testing. Symbol: n.

Acceptance region: In the theory of testing hypotheses, it refers to a region in the sample space such that, if a sample point falls within it, the hypothesis under test has been accepted.

Accumulated deviation: The term used for the graduation of a grouped frequency distribution provides, for each grouping interval, a 'theoretical' or 'expected' frequency. The difference of this quantity and the observed quantity taken with regard to sign and cumulated for increasing values of the varieties starting 'from the least, is termed as the accumulated deviation, It has been equivalent to the difference between the observed and theoretical distribution functions of the grouped distributions.

Accumulated process: Refers to a stochastic process which has been derived from the cumulated sum of random variables. For example, the total damage during time t of a sequence of thunderstorms is forming an accumulated process.

Accumulation point: An accumulation point of a set of points is a point P such that there is at least one point of the set distinct from P in any neighbourhood of the given point; a point which is the limit of a Sequence of points of the set (for -paces which satisfy the first axiom cd countability), An accumulation point of a sequence is a point P such that, for any integer n, each neighbourhood of P contains at least one term of the sequence after the nth term; e g , the sequence 1, ½. 1 1/3, 1, ¼, 1, 1/5, ... has two accumulation points, the numbers 0 and 1 Syn. cluster point, limit point. See Bolzano-Bolzano Weierstrass theorem, conden-sation-coride-nsation point.

Accuracy: In the general statistical sense this term denotes the closeness of computations or estimates to the exact or true values. In a more specialised sense the word is also having meaning: (a) in relation to an estimator, *unbiassedness* (see unbiassed estimator); (b) in relation to the reciprocal of the standard error, the precision. Neither usage can be recommended.

Adaptive inference: When we collect data over a period of time, the character of the inferences made at any one point will depend on the nature of previous observations and is then said to be adaptive. In a more restricted sense, the observations from a single may themselves determine the type of statistic used in obtaining inferences about the parent population.

Adaptive optimisation: The term used for a system of control for physical processes. It is based upon statistical adaptation us a series of observations proceeds through time. It is having a particular application in the field of statistical quality control.

Addition theorem: The principle holding that the probability that either or any one of two or more mutually exclusive events will occur in a given trial or occurrence is the sum of the individual probabilities.

Additive model: It is a model in which the factors influencing the dependent variable are having an additive effect.

Admissible decision function: In the general theory of statistical decision functions an admissible decision function may be defined as one for which it can be shown that there has been no decision function which has been uniformly better.

Admissible hypothesis: Generally, a hypothesis has been considered to be admissible when it has been possible within the conditions of the problem. More specifically, if there has been a distribution of known mathematical form depending on *k* unknown parameters θ then a statistical hypothesis could be stated by specifying any k set of θ s which has been a *priori* possible.

Admissible test: A test of a hypothesis is considered to be admissible if, in respect of a particular class of alternatives, there has been no other test with uniformly greater power e.g., if a uniformly most powerful test exists, no other test will be admissible.

Affinity: A concept which refers to a measure of divergence or distance between two statistical populations. It gives, for example, a generalisation of Mahalanobis distance.

Age: 1. Refers to the period of time that a person, animal or plant has lived, or, that an object has existed. 2. Also, refers to the level in year that a person has reached in any area of development compared with the normal level for his chronological age, for "ample, mental age, reading age.

Age-sex pyramid: A chart showing in one diagram the age distribution of males and females in a particular population. It is actually two histograms, placed back to back, with the bars lying horizontally, rather than standing vertically. A vertical line divides the chart into two parts". The left side of the chart, with its bars extending to the left from the vertical line, usually indicates the number or percentage of males in the particular age category represented by each bar. The right side, with bars extending to the Tight from the mid-line, provides the same information for females.

Age specific death rate: A term which is generally used in life analysis of human or biological populations for denoting the probability of dying in the next unit time period of (an entity that had survived for a stated period of) time. Substitution of word 'failure' for 'death' makes the term useful in connection with physical systems or components.

Aggregate: Refers to the value of a single variable which is results from the combination of data for a number of variables. For example an examination mark has been often an aggregate of the marks of several different papers, and perhaps a project. The process of forming an aggregate has been termed as aggregation.

Aggregation: A word which is

used to denote the compounding of primary data into an aggregate, generally for the purpose of expressing them in a summary form.

Aggregative index: Refers to an index-number which has been constructed by aggregating a number of items (as distinct, for example, from picking out a representative item). In price indices, if p_0 typifies the prices in the base period and p_n those for the current period and q_0 and q_n typify the quantities in the base and current periods respectively, then the two principal forms of aggregative index will be as follows:

$$\frac{\Sigma\,(p_n q_o)}{\Sigma\,(p_o q_o)} \text{ and } \frac{\Sigma\,(p_n q_n)}{\Sigma\,(p_o q_n)}$$

where the summation occurs over the commodities, to be aggregated.

Alias: This term finds use in two branches of statistical analysis. In connection with design of experiments, when a factorial design has been only fractionally replicated certain comparisons are not able to distinguish between some of the treatment combinations; these are considered to be aliases.

In connection with the spectrum analysis of time series an analogous concept arises since a single width of the interval of observation does not allow a distinction between certain angular frequencies.

Alienation: Refers to the effect of sampling errors on a correlation, coefficient.

Allocation, of a sample: Refers to the way in which sample numbers have been assigned to various parts of a population by the sampling plan; e. g., for a stratified population it could be decided to allocate the total sample number to the strata in proportion to the numbers of individuals in those strata.

Allowable defects: In quality control, the number of allowable defects in a sample may be equal to the critical number (designated by the particular scheme of sampling inspection) such that, if this number gets exceeded, the whole of the remainder of the inspection-lot should be examined or the lot rejected out of hand.

Almost certain: A probability dependent on a parameter n may tend to unity as n tends to infinity. The event to Which it relates is then considered to be 'almost certain' in the limit. More generally, it the measure of an event has been unity and its converse has measure zero but does not get related to an empty set, the event has-been 'almost' certain.

Almost everywhere: A statement about a value x in a domain is considered to be true almost everywhere with respect to a probability measure μ on x if the set of values of x for which the statement has been false has been of measure zero. Alternatively the statement is considered to be true almost surely, or almost certainly.

Almost stationary: A term which is used to cover the situation where the series resulting from wide interval sampling of a non-stationary time series has been itself stationary.

Alpha, error or a-error: In the theory of testing hypotheses it is an error incurred by rejecting an hypothesis when it has been actually true: more usually known as error of the first kind or Type I Error. In Quality Control it has been equivalent to producer's risk.

Alternating renewal process: Refers to a special case of the two-state semi-Markov process in which the two types of intervals are appearing alternately.

Alternative hypothesis: The conclusion we accept when the data fail to support the null hypothesis.

Amount of inspection: In quality control, is refers to the number of items (size of sample) taken from each lot and inspected in accordance to the particular sampling plan being used.

Analysis of dispersion: It is generalisation of the univariate analysis of variance providing a theory for multivariate tests of significance. It is sometimes termed as multivariate analysis of variance.

Analysis of variance or anova, *n*: It is a technique for investigating how much of the variability in a set of observations could be ascribed to different causes. In a one-way analysis of variance many different samples are drawn. The data would be then tested to see, if its variability (as measured by its variance) is all random or if parts of it is the result of systematic differences between the samples. Thus one-way analysis of variance tests the null hypothesis that all the samples have been drawn from the same parent population.

Analysis of variance has been has been involving separating the sample variance into two components, that within the samples and that between them. These are then compared using the. F-test. The variance is separated using the sum of squares result.

$$\sum x_T^2 = \sum x_B^2 + \sum x_W^2$$

Total sum of squares = That between the groups + That within them.

Analysis of variance may be carried out on the ranks of the data, rather than the actual values.

Analysis of variance for regression: The procedure for computing the F-ratio used to test the significance of the regression as a whole.

Analytic regression: Refers to a regression relation where the independent variables (regressors) are formed by polynomials, trigonometric sums or other analytic expressions.

Analytic survey: It is a (sample) survey where the primary purpose of the design has been the comparison between sectors or subgroup's of the population sampled.

Antilogarithm *n* antilogarithm of a given number: The number whose logarithm is the given number; e.g., $\text{antilog}_{10}\ 2=100$. *Syn.* inverse logarithm. To find an antilogarithm corresponding to a given logarithm that is not in the tables, subtract the next smaller mantissa from the given one and from the next larger one, divide the former difference by the latter and annex the quotient to the number corresponding to the smaller mantissa.

Antimode: Refers to the variate value, if any, for which a frequency distribution is having a minimum. The expression is generally confined to the case where the minimum, is not zero and is a true minimum, e.g. the zero tails of a Normal distribution at infinity are not antimodes in the sense.

Approximation: Refers to the process or result of making a rough calculation, estimate or guess. Approximation is used extensively when deciding if an answer has been reasonable. The statement 'It takes five hours to fly from Britain to Australia by jumbojet' is evidently, false. The distance is about 20,000, km, the speed about 1,000 km per hour, and so the time taken is about 20 hours. To decide that the statement was false, it was not necessary to know exact values of the speed of the aeroplane or the distance involved.

When numerical information is being given, the number of significant figures should be consistent with the accuracy of the information involved. Thus the statement 'The cost of running my car is Rs. 1.783 47 per year' is misleading because it is not possible to estimate it so accurately, especially when depreciation is taken into consideration. A better figure would have been approximately Rs. 1,800', which is obviously a round number, indicating an error of perhaps Rs. 50 either way.

Approximation error: Refers to an error due to approximation in numerical calculations as distinct, for example, from an error of observation. More particularly, it refers to a rounding error.

Area diagram: A type of graphic presentation in which a figure, such as a circle or rectangle, is divided into sections, each proportional in area in the size of the quantity it represents. The pie chart is a common type of area diagram.

Area sampling: A method of sampling used when no complete frame of reference is available. The total area under investigation is divided into small sub-areas which are sampled at random or by some restricted random, process. Each of the chosen sub-areas is then fully inspected and enumerated, and may from a frame for further sampling if desired. The term may also be used (but is not to be recommended) as meaning the sampling of a domain to determ-ined area, *e.g.*, under a crop.

Array: In the most general sense it refers to an explicit display of a set of observations. More usually, the term denotes some 'special arrangement of the observations, e.g., in order of magnitude. A frequency array may be defined as an array of frequencies accord-ing to variate values, that is to say, a frequency distribution. The term 'array' has been often used for the individual frequency distributions which form the separate rows and columns of a bivariate frequency table.

Artifact statistical: A statistical inference resulting from bias in the collection or treatment of data.

Association, statistical: Refers to a statistical relationship in which two or more variables occur or very together. Any of a variety of statistical techniques could be used to measure the extent to which a change in one variable results in a corresponding change in another variable or (variables). The selection of a particular technique is dependent upon such factors as whether the variables are qualitative or quantitative, the number of variables, and the nature of the data.

Assumed mean: Refers to an estimated or approximate value for the arithmetic mean, or average which is used to simplify its calculation.

When the arithmatic mean of a set of numbers of similar size is being found out, the calculation can sometimes be simplified by:

(a) Assuming a mean.

(b) Working out the differe-nces from it (+or–) for each number;

(c) Knowing the mean of the difference;

(d) Adding the mean of the differences into the assumed mean.

It does not matter what value is taken for the assumed mean, but the nearer it is to the true mean, the smaller are the numbers involved.

(a) Consider an assumed mean: 14y0m:

(b) Find out the differences from the assumed mean: —1, +4, —6, +6, +1 and +2 months;

(c) Find out the mean of the difference

$$\frac{-1+4-6+6+1+2}{6} = +1 \text{ month}$$

(d) The mean age of the children would be thus given by:

true mean = assumed mean + mean of differences = 14y0 + 1m

The mean age of the children would be 14ylm.

Assumed mean: An arbitrary origin or working mean which is used for the calculation of moments. The expression has not been to be recommended.

Asymmetrical distribution: Refers to a distribution which is not symmetrical, that is to say, for which there is no central value a such that $f(x—a) = f(a—x)$, $f(x)$ being the frequency function.

Asymptotic distribution: Refers to the limiting form of frequency or probability distribution which is dependent on a parameter, such as sample-number or time, as that parameter tends to limit, usually infinity.

Asymptotic normality: A distribution dependent on a parameter n, generally a sample number, is considered to be asymptotically Normal if, as n tends to infinity, the distribution tends to the Normal form.

Asymptotic standard error: The standard error of a statistic nearly always depend on the sample number n. If, as n tends to infinity, the standard error S is asymptotically equivalent to a quantity Q, *i.e.* is such that S/Q tends to unity, then Q may be termed the asymptotic standard error. As so defined, Q is, not unique but in the vast majority of cases S^2 is expressible as a term in n^{-1} plus terms of lower order in n, and the first is taken as the square of Q.

Attenuation: Where observations on bivariate material have been subject to errors of measurement the true correlation between the variates will be obscured, usually being underestimated. The correlation is then considered to be attenuated. In 1904 Spearman advanced the following formula which would correct this attenuation

$$r'_{xy} = -\frac{r_{xy}}{\{r_{xx}\, r_{yy}\}1/2}$$

where r_{xy} denotes the geometric mean of correlations between independent determin-ations of x and y; r_{xx} and r_{yy} have been the means of correlation between independent determina-tions of x

and of y and $r'_{xy'}$ is the corrected correlation.

Attenuation, statistical: The reduction in the size of a correlation through errors of measurement. Because of errors the correlation obtained in a study is lower than the true correlation that exists in the population. A correction for attenuation may be used to estimate the size of the correlation without the chance errors. The obtained correlation is divided by the square root of the product of the reliability coefficient; of the two variables. The resulting value is the maximum probable correlation between the variables. It should be remembered, however, that this is a theoretical and not an obtained value.

Attribute: An-inherent property, usually one that is measured by a finite number or a discrete set of values. Sometimes used as the outcome of ore of a set of mutually exclusive events; when used in this sense, attribute space is the same as sample space. Often attribute, sampling refers to an experiment with only two outcomes, e.g., classifying an object as defective or nondefective.

Attribute testing: Refer to the testing of a product for quality control when the result has been either 'good' or 'defective' Attribute testing could be contrasted with variable testing where the result is a quantitative measure, like the length of a nail or the resistance of an electrical component.

Autocorrelation: Refers to the internal correlation between members of series of observations ordered in time or space.

Autocorrelation, function: The autocorrelation function of a stationary stochastic process may be defined as the autocovariance divided by the variance.

Autonomaus equations: In econometrics, refers to an equation which is describing the behaviour of one particular group or sector of the economy, e.g.. a demand equation describing only buyer's behaviour, is regarded to be autonomous if it is affected only by changes in the behaviour of that particular group or sector. Equations of this kind considered sometimes termed 'structural equations' but this is not be, recommended, as such equations may include variates from the whole system.

Autoregression: Refers to the generation of a series of observations where by the value of each observation has been partly dependent upon the value of those which have immediately, preceded it, *i.e.,* each observation stands, in a regression relationship with one or more of the immediately preceding terms. A scheme of autoregression may 'be regarded to be as a stochastic process of a conditional kind.

Autoregressive model: Refers to an econometric model which is based upon, the autoregressive process but also containing lagged versions of some, or all of the endogenous variates appearing in the model specification.

Autoregressive series: Refers to a series which was generated by an autoregressive process. It is the realisation of an autoregressive process.

Autoregressive transformation: If there has been autocorrelation in the error term of an autoregressive process it has been sometimes possible to, transform the original variates to new variates such that the autoregressive scheme in the transformed variates is having an uncorrelated error term. This procedure is termed as an autoregressive transformation.

Average: Refers to a numerical value that is a central point about which other values in a series get dispersed. An average provides a single, summary measure of a distribution of numerical values, In statistical terms it is usually referred to as a measure of central tendency, for it summarizes in one value the central tendency of all the variations in a distribution taken as a whole. There have been different types of averages, or different measures of central tendency, each of which way result in a different average value for the same series of values.

Average deviation: A synonym of mean deviation, not to be recommended. It is sometimes called the mean deviation, it is a simple measure of dispersion of a series of figures from their arithmetic mean. It is equal to the sum of the deviation from the mean, with sign ignored, divided by the number of figures, in the series.

$$\text{Average deviation} = \frac{\Sigma x}{n}$$

where Σ is the sign of summation, x the deviation from the mean, and n the number of figures in the series. Statisticians prefer the standard deviation. It is having wide and general application in all fields except statistical theory.

Average of relatives: An average which is usually in the form of an index number, of a set of relatives, that is to say values obtained as the ratio of a magnitude in the given period to the corresponding magnitude in the base period. In price index numbers, the 'price relatives are generally weighted by the values either of the base period or of the given period. Where the weights used have been the values in the base year the formula reduces to that of Laspeyres. If values in the given period have been tied with a harmonic average the formula reduces to that of Paasche.

Average Outgoing Quality (AOQ): Refers to the proportion of defective items being sent out with a product after a quality control scheme has been implemented. For a single inspection scheme (single sampling scheme) with acceptance number n from samples of size N, with 100% inspection of failed batches AOQ may be put as follows :

$$AOQ = P(p)\ (1—n/N)p$$

where, p refers to the proportion of defectives manufactured, and P(p) refers to the probability of a batch with proportion p defective being accepted.

Average Outgoing Quality Limit (AOQL): Refers to the highest average proportion of defective items that will be let through in the long run, following a quality-control sampling scheme.

When we plot average outgoing quality (AOQ) against the proportion of defectives manufactured, p the graph rises to a maximum and then falls away. A typical graph for a single sampling scheme has been defected in Fig. If p is small, the AOQ will be low because there are few defectives' to get through. If p is large, the sampling scheme is likely to result in most batches being stopped and undergoing 100% inspection; any defectives are then rejected. Between these two cases have been values of p which are neither very small, nor large enough to involve most batches being stopped.

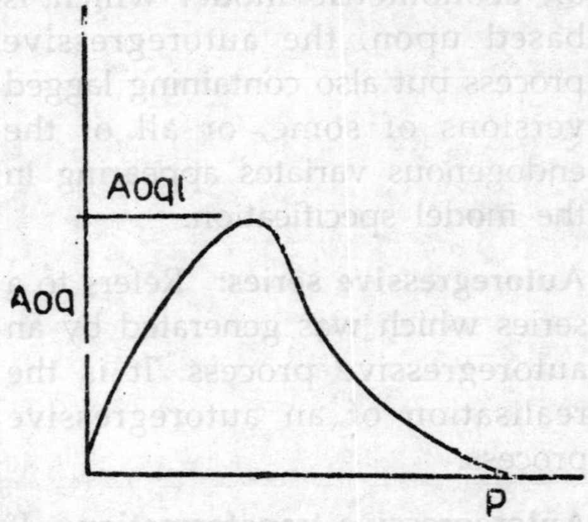

Average outgoing quality limit (AOQ). The average outgoing quality (AOQ) of goods is plotted against the proportion of defective goods (p) to find the maximum value (AOQL).

The maximum value shown on such a graph is known as the average outgoing quality limit, and represents the highest average proportion of defectives that will be let through in the long tun. It has been however, possible for worse individual batches to get through.

Average Run Length (ARL): The average run length of a sampling inspection scheme at a given level of quality may be defined as the average number of samples of n items taken in the period between the time when the process commences to run at the stated level and that at which the scheme reveals a change from acceptable to rejectable quality level is likely to have occurred. This concept must not get confused with average article run length.

Average sample number curve: Refers to the graph of the average sample number function, with the function as ordinate against the parameter as abscissa.

Average sample run length: An alternative name which is sometimes used for average run length so as to distinguish it from the concept of average article run length.

Axial distribution: It is a spherical distribution which describes the relative frequencies of the possible orientations in three dimensions of a random a axis. As an axis in a direction without a sense, such a distribution can be defined entirely on a unit hemisphere, or conventionally on a unit sphere by assigning the same probability density to diametrically opposite points.

B

Balanced differences: In a systematic sample of an ordered series the selected units are not located at random and therefore no fully valid estimate of sampling error would be possible in general. One method of overcoming this difficulty has been to construct artificial strata by dividing the series into 'blocks' of equal length and to consider the members falling within a block as having been-chosen at random within that block.

If there happen to be only two members in the block an estimate of error will be based on their difference. If there happen to be more it may be based on more complicated linear functions designed to eliminate systematic error, e.g., for seven members there would be used 'balanced differences' of the form.

$d = ½y_1 - y_2 - y_3 - y_4 + y_5 - y_6 + ½y_7.$

The numbers of terms has been arbitrary but seven or nine will eliminate most of any systematic component of vari-ation. Similar ideas have been found to be applicable to syste-matic sampling in more than one dimension.

Balanced sample: When the mean value of some characteristic is known for a population and the value of the characteristic can be ascertained for each member of a sample, it becomes possible to choose the sample so that the mean value of the characteristic in it approximates to the parent mean. Such a sample is considered to be balanced.

The object of balancing is to get a sample which is representative of the parent in respect of some other characteristic for which the parental value is not known. Authorities have been reported to differ about the value and validity of the method.

Ballet theory: Refers to the various generalisations and extensions of a theorem by Bertrand: when a candidate (A) scores a votes and candidate (B) scores (b) votes, a>μb and μ is non-negative integer, then the probability throughout the counting that the votes for (A) are always greater than μ times the votes for (B) will be Pr = $(a-\mu b)/(a + b)$.

Bar chart or bar diagram: A display of data using a number of rectangles, of the same width, whose lengths are proportional to the frequencies they represent. Bar charts can be drawn in several ways; for example, with the bars drawn vertically or horizontally immediately next to each other, or with, a fixed gap between them. A component or compound bar Chart has two or more parts to each bar. There are many ways in which such diagrams can be drawn, some of them very artistic. The lengths of the bars and the areas of the constituent parts are pro rata to the absolute magnitudes of the statistics that they represent.

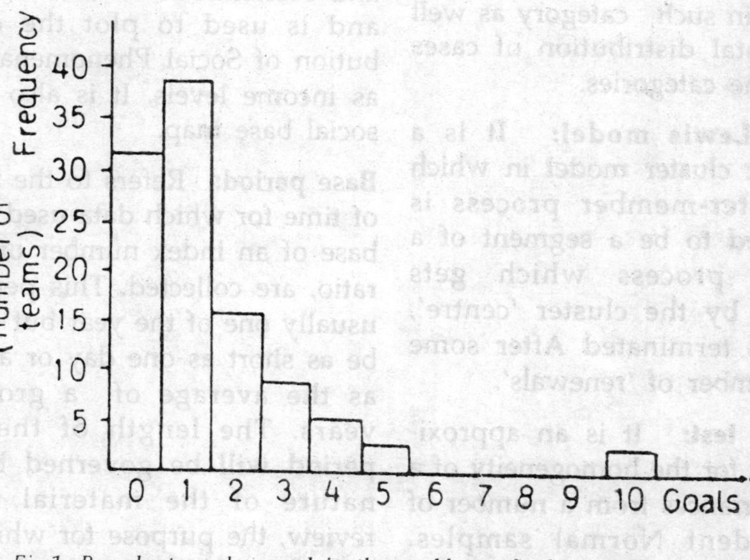

Fig.1. Bar chart, goals scored in the world cup finals in 1982 in Spain

Bar graph: A type of graphic presentation of data in which either horizontal or vertical bars are drawn to represent a distribution of cases. The bars are drawn to scale and represent proportionately the quantities being compared. The bar graph is most often used to portray the frequency distribution of a qualitative variable, with each bar representing a category of the variable, its height or length indicating the number of cases in that category. Bar graphs are also used to portray time series, with the height or length of each bar indicating the frequency or magnitude of the variable in each time period. They may also be used with quantitative variables, with the height or length indicating the number of cases in each class interval.

Bar graph, divided: A bar graph in which the bars are divided into segments that show the number of cases in two or more subdivisions of each category. For example, if the bars in a graph represented the categories of the variable marital status each might be divided into two segments representing males and females, thus showing the proportion of each sex in such category as well as the total distribution of cases among the categories.

Bartlett-Lewis model: It is a stochastic cluster model in which the cluster-member process is considered to be a segment of a renewal process which gets initiated by the cluster 'centre', and gets terminated After some finite number of 'renewals'.

Bartlett's test: It is an approximate test for the homogeneity of a set of variances from a number of independent Normal samples. This test was given by Bartlett in 1937.

Base: A number or magnitude which is used as a standard of reference, It may be found as a denominator in a ratio or percentage calculation. It may also be the magnitude of a particular time series from which a start has to be made in the calculation of a new relative series-an index number which will show the observations as they get accrued in the future in relation to that of the base period.

Base line: It is the horizontal line on a graph which is corresponding to some convenient basic measurement of the variable represented on the ordinate scale. The base is generally taken to be zero.

Base map: A map that shows the location of Important geographic and economic features of an area, and is used to plot the distribution of Social Phenomena, such as income levels. It is also called social base map.

Base period: Refers to the period of time for which data used as the base of an index number or other ratio, are collected. This period is usually one of the year but it may be as short as one day or as long as the average of a group of years. The length of the base period will be governed by the nature of the material under review, the purpose for which the index number (or ratio) is being compiled and the desire to use a period as free as Possible from abnormal influences so as to avoid bias.

Base weight: Refers to the weights of a weighting system for an index number which are computed according to the information relating to the base period instead, for example, of the current period

It is general in writing formulae to denote the information from, the base period with a suffix o and that for the given, Period with a Suffix n. A price index which is weighted according to Prices and quantities of the base, i.e., base-weights, might be

$$I_{on} = \frac{\sum p_o q_o \left(\frac{P_n}{P_o}\right)}{\sum p_o q_o}$$

where p_o, q_o denote the base weights and summation is taken over the commodities composing the index.

Basic statistics: The raw materials of statisticians; usually Collated, but rarely processed beyond the stage of correction for extraneous factors, such as differences in Materials prices and wage rates where the basic statistics consist of annual costs. They can be classified in various ways, one being the dichotomy of continuous series and discrete series; another dichotomy is that of time series and cross section series and a third, that of internal and environmental statistics.

Battery tests: A term which is used in applied psychology for a group of tests to which subjects get submitted. The usual objects of subsequent analysis have been either provide prediction of each subject's aptitude for one or more occupations on the basis of a weighted combination of series in the tests, or else to provide variables Whose correlations may sub-sequently get analysed into group or general factors common to the tests.

Bayes' estimation: Refers to the estimation of population parameters by the use of methods of inverse probability and in particular of Bayes' theorem. If Pr(θ/H) represents the prior probability of q then the posterior probability of q will

$$\Pr(\theta/x_1, x_2, \ldots, x_n, H) =$$

$$\Pr(\theta/H) \Pr(x_1, x_2 \ldots, x_n/\theta H)$$

θ is estimated by selecting that value which maximises the Posterior Probability. If Bayes' postulate gets invoked Pr(θ/H) is Constant and the method will be equivalent to the maximisation of the likelihood $\Pr(x_1, x_2, \ldots x_n/\theta H)$.

Bayesian controversy: Refers to a controversy between two schools of statistical thought. In Bayesian statistics, Probability is regarded as a numerical degree of belief. In some situations, like throwing a pair of dice, the belief can be supported objectively. In others such as predicting the price of gold next week, any probability used has been essentially subjective. In non-Bayesian statistics, probabilities must rely on a frequency based interpretation.

Statistical inference has been involving drawing conclusions about the situation underlying experimental, data, and this may involve the use of Bayes' theorem on conditional probability.

The controversy arises because a non-Bayesian statistician fails to accept the validity of using this theorem if the probabilities are not frequency based.

Bayes' postulate: Refers to a Postulate which is concerning the prior probabilities of a set of hypotheses (of Bayes' theorem) to the general effect that in the absence of information to the contrary all prior probabilities are to be considered equal. The Postulate (or something equivalent) has been necessary if Bayes' theorem has to be applied in situations where there has been no information concerning the prior probabilities.

Bayes' risk: Refers to the minimised value of the expected risk in a Bayes' solution.

Bayes solution: In the terminology of statistical decision functions, a Bayes' solution refers to a decision function which is able to minimise the average risk relative to some probability distribution.

Bayes' strategy: Refers to one for which the available decision rules could be linearly ordered in terms of the Bayes' risk.

Bayes' theorem: (On conditional probability), the theorem which states that

$$P(A \mid B) = \frac{p(B|A)\, p(A)}{p(B|A)\, p(A) + p(B|A')\, p(A')}$$

where $p\ (A \mid B)$ refers to the probability of event A occur i given that event B has occurred, and where A' is the event 'not' A'. Hence Bayes' theorem relates p(A | B) to p(B | A) and p(B | A').

Bayes' theorem can be written in more general form as

$$p(A_r \mid B) = \frac{p(B|A_r) \times p(A_r)}{\sum_{r=1}^{r=n} p(B - A_r) \times p(A_r)}$$

where A_1 ... A_r are mutually Exclusive and Exhaustive events. The theorem enables the probabilities of the explaining hypotheses to be determined; this use is called the method of inverse probability.

The principal difficulty lies in determining, or even defining, the prior probabilities and the resolution of the difficulty by Bayes' postulate has occasioned much controversy. This does not, however, affect the theorem, which is a simple consequence of the product law of probability.

Bayesian inference: Refers to a

type of inference which regards parameters as being random variables possessed of prior distributions reflecting the accumulated state of knowledge. In its methods it is making use of Bayes' theorem and its extensions.

Bayesian intervals: It is a method of summarising the information which is contained in the posterior probability distribution by quoting one or more intervals which contain certain amounts of probability. When such intervals have been based on a non-informative prior probability distribution they are sometimes termed as credible intervals or Bayesian confidence intervals. When the posterior distribution has been symmetric, i.e., that the tails have equal densities, then these intervals are termed as highest posterior density intervals because every point included is having a higher probability than these excluded.

Bayesian probability point: It is a critical value which is analogous to a confidence point or limit and derived according to the Philosophy of Bayes' theorem and the use of the prior distribution (*s*) of the parameter (*s*) under estimation.

Behavioural decision function: In statistical decision theory, it is a rule which is used for choosing between the possible decision functions, based on the outcome of an experiment.

Bell-shaped curve: It is a symmetrical frequency curve, which is usually of a continuous frequency distribution, and shows it marked similarly to a vertical section through a bell.

Bellman-Harris, process: It is a branching process which evolves from an initial individual, in which each individual lives for a random length of time and at the end of its life produces a random number of offsprings of the same type.

Belt graph: A type of segmented line graph in which several line graphs, combined in a single chart, show the component elements being depicted as well as their total. The belt graph is the linear counterpart of the divided bar graph. It shows the distribution of the component parts individually and together.

Bernoulli distribution: It is another name for Binomial Distribution. The term is, however, generally restricted to the special case of the binomial distribution when $n = 1$

Number of successes:	0	1
Probability:	$1-p$	p

Thus it is the probability distribution of the number of successes in a single trial.

Bernoulli numbers: The Bernoulli number of order r, B_r, is defined a's the numerical coefficient of $t^r/r!$ in the expansion of $(e^t - 1)$ as a power series in x.

Explicitly, $B_0 = 1$, $B_1 = -½$, $B_2 = 1/6$ $B_4 = -\frac{1}{30}$, $B_6 = \frac{1}{42}$, $B_3 = B_5 = B_7 = \ldots = 0$.

There are variations in the notation and some writers give

$$B_1 = -½,\ B_2 = 1/6,\ B_3 = \frac{1}{30} \ldots$$

Bernoulli polynomial: The Bernoulli polynomial $Br^{(n)}(x)$, of order n and degree r is defined as the coefficient of $t^r/t!$ in the expansion of

$$\left(\frac{1}{e^t - 1}\right)^n e^{xt}.$$

Bernoulli process: A process in which each trial has only two possible outcomes, the probability of the outcome of any trial remains fixed over time and trials are statistically independent.

Bernoulli's theorem: A theorem propounded by James Bernoulli in fourth part of his Ars Conjectandi which was published in 1713 after his death in 1705. It is the theorem which states that if the probability of success in a trial is p, the probability of exactly r successes in n independent trials is.

$$nC_r p_r\ (1- p)^{n-r}$$

where $nC_r \frac{n!}{(n-r)!r!}$ and $n!$ (n factorial) is

$$1 \times 2 \times 3 \times \ldots \times n$$

This is the theorem underlying the Binomial Distribution.

Effectively it is a proposition in pure mathematics to the effect that the observed proportional frequency in random drawings of individuals from a population of attributes with constant probability p converges to p in probability; or, to put it another way, the proportional frequency in the binomial distribution $(q+p)^n$ lying within a range $\pm\varepsilon\sigma$ of the mean value p tends to unity with increasing n however small ε may be, σ being the standard deviation $\sqrt{(pq/n)}$.

Bernoulli himself seems to have regarded the theorem as something beyond a mathematical proposition, perhaps a justification of a frequency theory of probability.

Bernoulli trial: It is an experiment with fixed probability p of success, $1-p$ of failure. In a sequence of 'Bernoulli trials, the probability of success remains constant from one trial to the next, independent of the outcome of previous trials.

The probability distribution for the various possible numbers of successes, 0 to n, in n Bernoulli trials has been the binomial Distribution. This can also be called the bernoulli distribution, but this term is often restricted to the case when $n = 1$.

Bernoulli variation: Refers to a sampling situation in which members are selected from a population of attributes such that the probability of occurrence remains constant, hence the sampling distribution of occurrence in samples of fixed size has been binomial. The term finds use in contradistinction to Lexis and Poisson variation.

Bessel function distribution: It is a frequency distribution which is involving Bessel functions. For example, the distribution of the covariance of two Normal correlated variates involves a Bessel function of the second kind having imaginary argument.

Bessel's correlation: A factor by which the variance of a small Sample is multiplied to provide an estimate of the variance of the universe. The correction factor is

$$\frac{n}{n-1}$$

where n is the number of items in the sample. The greater the value of n. The closer the correction approaches to unity, so that for very large samples the correction becomes unnecessary.

Best estimator: The estimation of population parameters from information provided by the sample gives rise to the question whether there has been a 'best' estimator. The answer has been found to depend mainly on the criteria which are laid down as to the 'goodness' of an estimator. If there has been a criterion which distinguishes one of two estimators as better than the other and if there exists an estimator which is better than other, it is considered to be the best.

Various criteria are suggested, e.g. that of sufficiency, minimum variance or closeness. It is not always true that a 'best' estimator exists.

Best fit line of: A straight line or curve fitted as a graph representing the relationship between two variables plotted one against the other in a scatter diagram. Figure shows a scatter diagram of monthly costs plotted against monthly output, The line of best fit may be drawn freehand by personal judgement, as it is in Figure; or it may be fitted by group averages or least squares, or by any other method of simple-regression analysis. When the line has been drawn, then if it turns out to be straight, readings can be taken from it in order to derive its equation. If it turns out to be a curve, then it would be necessary for analytical purposes to determine the type of equation to which it conforms, e.g. logarithmic or quadratic. See model building.

Best linear unbiassed estimator. It is a linear function of the order statistics which is available in the sample of information of the variable under analysis for which the variance has been minimum and which has been also Unbiassed.

Beta-binomial distribution: Refers to a discrete probability distribution of the form given below:

$$p(r) = \binom{n}{r} \frac{B(\alpha + r, n + \beta - r)}{n(\alpha, \beta)} \quad r = 0, 1, 2, \ldots n.$$

It is the convoluted distribution which is obtained by averaging the probability parameter of a binomial distribution having index n over a beta distribution having parameters α and β.

Beta coefficient (β). A standardized measure of a regression coefficient (R8); achieved by 'stating each of the variables in units of their own individual standard deviation'. A simple regression equation (R9) may be written

$$Y = aX + b$$

Its beta equivalent is:

$$\frac{Y}{\sigma y} = \beta \frac{X}{\sigma x} + b^1$$

Each figure of the Y series is divided by the standard deviation of the Y series, and each figure of the X series is divided by the standard deviation of the X series, before the regression analysis is carried out. The standard deviation of a series expressed in units of cwt is 20 times that of the same series expressed in units of a ton; the resultant ratios are the same whether the series is expressed in cwts or tons. At the same time the size of any regression coefficient depends partly on the Units in which the basic series are expressed. The beta coefficient eliminates this factor from the regression, and the several beta coefficients of a multiple-regression equation can be compared directly one with another, and with the beta coefficients, of other regression equations.

Beta distribution: Refers to the distribution with probability density function given by:

$$f(x) = \frac{x^{\alpha-1}\,(1-x)^{\beta-1}}{B(a\beta)}$$

where the parameters αβ>1, $0 \le x \le 1$, and B(α, β) is the beta function given by

$$B(\alpha, \beta) = \int_0^1 x^{a-1}\,(1-x)^{\beta-1}\,dx$$

The beta distribution is having

$$\text{Mean} = \frac{\alpha}{(\alpha + \beta)}$$

$$\text{Variance} = \frac{\alpha\beta}{(\alpha+\beta+1)\ (\alpha+\beta)}$$

The graph of the beta distribution get varied in shape according to the values of α and β as shown in, Fig. 1.

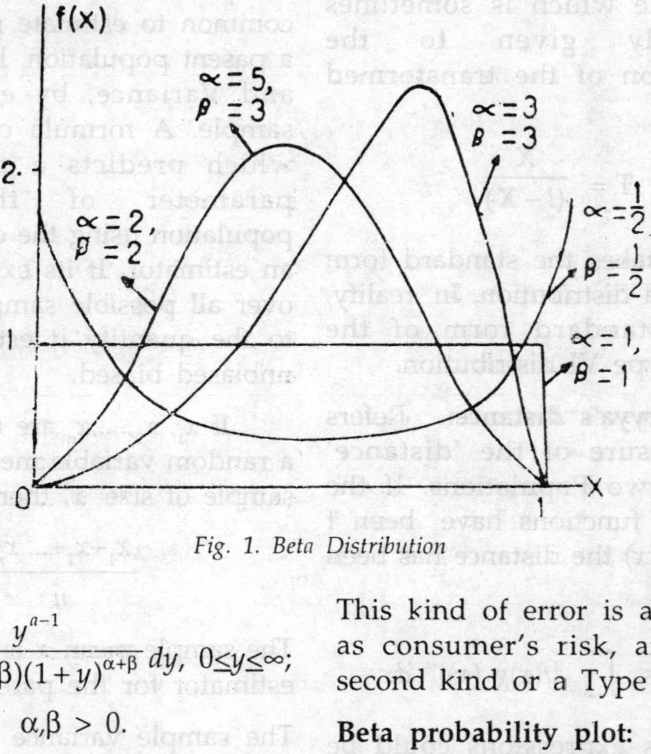

Fig. 1. Beta Distribution

$$dF = \frac{y^{a-1}}{B(\alpha,\beta)(1+y)^{\alpha+\beta}}\,dy,\ 0 \le y \le \infty;$$

$$\alpha, \beta > 0.$$

This easily gets transformed into the first type by putting $x = 1/(1 + y)$ It is also referred to is an 'inverted' beta distribution.

Distributions of the first kind have been a special case of the Pearson type I distribution and those of the second kind Fire a special case of the Pearson type VI distribution.

β-Error. In the theory of testing hypotheses, it refers, to an error which is incurred by accepting an hypothesis when it is incorrect. This kind of error is also termed as consumer's risk, an error of second kind or a Type II error.

Beta probability plot: Refers to the graph of cumulated distribution function as ordinate against the variate as abscissa for beta distributions. For Type I the variate is having finite range. For Type VI it is infinite and is used to test a variance ratio. It does not involve specially ruled paper. The plots may be of two types: (a) ordered quantities in the unit range against quantities of the appropriate beta 'distribution of the first kind (Pearson Type 1) or (b) ordered rations of mean squaıes against quantities of the appropriate beta distribution of the second kind (Pearson Type VI).

Beta-prime distribution: Refers to a name which is sometimes regrettably given to the distribution of the transformed variable

$$T = \frac{X}{(l - X)}$$

where X takes the standard form of the beta distribution. In reality, it is a standard form of the Pearson type VI distribution.

Bhattacharyya's distance: Refers to a measure of the 'distance' between two Populations. If the frequency functions have been f (x), and g(x) the distance has been given by

$$\text{arc cos} \int_{-\infty}^{\infty} \{f(x)g\ (x)\}^{1/2}\ dx$$

Analogues expressions could be given for discontinuous and multivariate distributions.

Bias: Refers to a systematic error in a statistical result. Error's from chance will cancel each other out in the long run, but those from bias will not.

Generally, bias has been an effect which deprives a statistical result of representativeness by systematically distoring it, as distinct from a random error which may distort on any one occasion but balances out on the average.

Biased estimator: A formula or procedure which produces a biased estimate of a Parameter of the parent population from sample data. In statistics it is common to estimate properties of a pasent population, like its Mean and Variance, by examining a sample. A formula or procedure which predicts a property or parameter of the parent population using the data is called an estimator. If its expected value over all possible samples is equal to the quantity it estimates, it is unbiased biased.

If $x_1, x_2, ..., x_n$ are the values of a random variable measured, for a sample of size x, then

$$\frac{x_1 + x_2 + ...x_n}{n} = \bar{x}$$

The sample mean x is an unbiased estimator for the parent mean.

The sample variance is, however, a biased estimator for the parent variance:

$$\sum_{i=1}^{n} \frac{(x_1 - \bar{x})^2}{n}$$

An unbiased estimator for it is:

$$\sum_{i=1}^{n} \frac{(x_1 - \bar{x})^2}{n - 1}$$

Also, see unbiased estimator.

Biased sample: Refers to a sample which is obtained by a biased sampling process, that is to say a process which incorporates a systematic component of error, as distinct from random error which balances out on the average. Non-random sampling is generally, though not inevitably, subject to bias, particularly when entrusted to subjective judgement on the part of human beings.

Biased test: A test is considered to be biased if it provides a lower probability of rejecting the hypothesis under test (H_0) when the alternatives hypothesis (H_1) has been true then when H0 has been true. Expressed in another way if the hypothesis under test has been $\theta = \theta_0$ and the power function of the test has a minimum value at a point $\theta \neq \theta_0$ then the test has been considered to be biased.

Bimodal (of a distribution): 1. Having two Modes. For example, the goals scored per match by a football team in a season are given in the table below. As there have been two modes, 0, and 2, so the distribution has been bimodal.

Goals/match	*Frequency (matches)*
0	12
1	8
2	12
3	6
4	3
5	1

2. A distribution having two distinct, but unequal, local modes, a shown below:

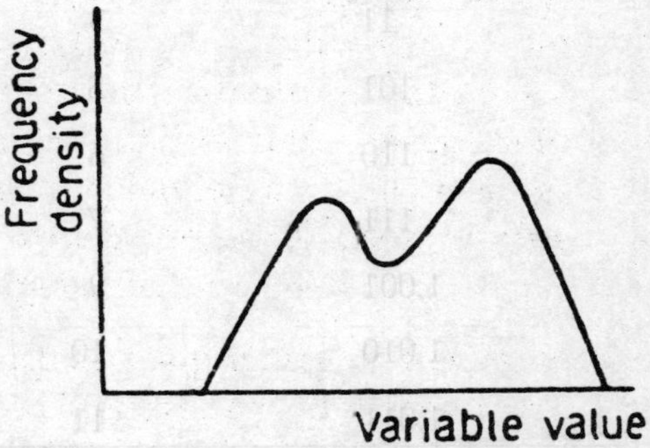

A bimodal distribution.

Bimodal distribution: Refers to a distribution of data points in which two values occur more frequently than the rest of the values in the data set.

Binary numbers: Decimal numbers consist of the ten numbers .0 to 9; binary numbers, of the two numbers 0 and 1. In binary, then, 10 equals 2 of the decimal notation, and is described in terms of the decimal system as being of the base 2; the decimal system is described in its own terms as being of base 10. The following table of equivalents is a useful and revealing guide to conversion.

Binary	*Decimal*
1	1
10	2
100	4
1,000	8
100,00	16
100,000	32

By addition:

11	3
101	5
110	6
111	7
1,001	9
1,010	10
1,011	11

It will be seen that, as the binary numbers shown proceed from 1 upwards in powers of 10 (binary), the corresponding decimal numbers proceed in powers of 2 (decimal), i.e., 2^0, 2^1, 2^2,2^3,...

Computers use the binary system internally for technical reasons. Print-outs are invariably stated in decimal numbers.

Binary sequence: In general, it refers to any sequence each numbers of which can take one of two possible values. In a probabilistic context it refer to a series of numbers each of which represents the outcome of an (independent or dependent) binomial trial and hence where the two possible out-comes might be represented as 0 or 1.

Binomial coefficients: Refers to the coefficients of the terms in the expansion of $(x+y)^n$ for positive integer values of n. Symbol: ${}^nC_r$.

Binomial distribution: The theoretically expected distribution of a series of random samples drawn from a population with two categories or classes. A binomial distribution would show the theoretical frequency with which one would expect to obtain random samples with each of a series of proportions of cases in the one of the two categories. For example, if 50 random samples were drawn from a Population half male and half female, binomial distribution could give the theoretically expected number of samples that would be all made, 90% male, 80% male, 70% male, etc. The binomial distribution is one of the important mathematical models used in statistics, See poisson distribution; sample, random.

Binomial, distribution is also known as the Bernoulli distribution after James Bernoulli who give it in his (posthumous) Ars Conjectandi in 1713.

If the probability of success is p, and that of failure $q = 1- p$, then the probability of r successes and $n\text{-}r$ failures is given by:

${}^nC_r p^r q^{n-r}$ (the binomial coefficient

$${}^nC_r \left(\frac{n!}{r!(n-r)!} \right)$$

The probability generating; function for the binomial distribution has been given by $(q+pt)^n$ since, when expanded, this provides;

$${}^nC_0 q^n + {}^nC_1 q^{n-1}\ (pt) + \ ...\ + {}^nCrq^{n-r}(pt)r + \ ...{}^nC_n\ (pt)^n$$

It is often convenient to write the

expression to be expanded as $(pX + pY)^n$, where event X is having probability p, event Y, i.e., X', or not X, has probability q = 1 - p. X and Y can then be thought of as markers, so that the probability of X occurring r times, and Y occurring n-r times, is given by the coefficient of X^rY^{n-r} namely:

$$^nC_{n-r}p^rq^{n-r} = \frac{n!}{(n-r)!r!}\,p^r q^{n-r}$$

The mean, or expectation, of the binomial distribution is given by np, the standard deviation by $\sqrt{npq}$.

The distribution of the proportion of successes in n trials is binomial with

$$\text{Mean} = p$$

$$\text{Standard deviation} = \sqrt{\frac{pq}{n}}$$

Binomial index of dispersion: A statistic which is used for testing Whether a set of samples has been homogenous with respect to some common attribute.

If there have been k samples of sizes $n_1,\ldots n_k$ with proportion $p_1,\ldots,p_k$ and p has been the mean proportion for all members together; ie.,

$$p = \left(\sum_{i=1}^{k} n_i p_i\right) / \sum_{i=1}^{k} n_i,$$

the index of dispersion is

$$\sum_{i=1}^{k} n_i\ (p_i - p)^2\{p(1-p)\}.$$

The significance of the index, as denoting departure from, homogeneity, may be tested, in the χ^2 distribution having k-1 degrees of freedom.

The index has been a particular case of the Lexis ratio.

Binomial probability paper: A graph paper having a gird which is specially designed to facilitate the analysis of enumeration data, i.e., data in the form of proportions or percentage from a binomial population. Both the rectangular coordinate axes have been graduated in terms of the square-root of the variable.

Binomial test: A nonparametric statistical test that may be used when research data are divided into just two categories (e.g., male and female, favourable responses and unfavourable responses, over 21 years of age and under 21 years of age). The binomial test uses the Binomial Distribution to test whether the distribution of cases in the sample studied differs significantly from the distribution that would be expected if the sample were drawn randomly from a particular two-class population. It is a Goodness-of-Fit Test.

Binomial variation: It is another name for Bernoulli variation.

Binomial waiting time distribution: It is alternative name for the negative binomial distribution which refers to the realisation of a success for the kth time after $k+x-1$ independent binomial trials.

Biometry: Refers to the study of biological data by means of statistical analysis.

Birth and death process: It is a stochastic process which tries to describe the growth and decay of a population the members of which may die or give birth to new individuals. The types mainly studied passes relatively simple laws of reproduction and mortality.

Birth, death and immigration process: It is a simple extension of birth and death process which takes into consideration immigr-ation into the system out-side. Immigration could be treated synonymously with 'death'.

Birth process: It is a stochastic process which describes the population of a system in which individual members may give birth to new members. The expression is generally confined to the case where the variate (population) increases only by jumps of amount + 1, the probability of a jump from n to n +1 in time dt being asymptotically $\lambda_n$dt. Here λ_n may also be depending on t.

Birth rate: The crude birth rate of an area may be defined as the number of births actually occurring in that area in a given ,time period divided by the population of the area as estimated at the middle of the particular time period. The rate has been usually expressed in terms of 'per 1000 of population'.

This crude birth rate has been capable of considerable refinement according to the various specific viewpoints adopted in the analysis of vital statistics. For example, it May get adjusted to allow for changes 'in the proportion of the female population in the child-bearing age groups.

Biserial: (Of the relationship between two variables) having one variables that takes only two values.

Bit: An abbreviation for binary digit. It is a digit of a number which is written in the scale of two; for example, the number 2 is expressed as 10, 4 as 100 and 8 as 1000.

Bivariate beta distribution: It is a bivariate continuous distribution which is due to Filon and Isserlis , having the joint probability density

$$h(x,\ y)=\frac{\lambda(p_1+p_2+p_3)}{\lambda(p_1)\lambda(p_2)\lambda(p_3)}\,x^{P1-1}y^{P2-1}$$

$$(1-x-y)^{p3-1}$$

where $x \geq 0$, $y \geq 0$, $x+y \leq 1$ and p_1, p_2, $p_3 > 0$. The distribution can be derived as the joint distribution of

$$X = X_1/(X_1+X_2+X_3),$$
$$Y = X_2/(X_1+X_2+X_3)$$

where X_1, X_2, X_3 denotes, independent random variables having gamma distributions with index parameters p_1, p_2, and p_3.

Bivariate binomial distribution: It is an extension of the binomial distribution to the case where a member can exhibit success or failure in each of two attributes.

Bivariate cauchy distribution: A bivariate probability distribution having density function

$$f(x_1, x_2) = \frac{a}{2\pi(a^2+x_1^2+x_2^2)^{3/2}}, -$$
$$\infty < x_1, x_2 < \infty, a > 0$$

Bivariate data: Data in which there are two measurements for each item, for example the income and years of education of each subject studied.

Bivariate distribution: Refers to the distribution of a pair of variates x_1 and x_2, written for the continuous case as

$$dF = f(x_1, x_2)\, dx_1, dx_2.$$

For discontinuous or grouped data the distribution may be set cut in a rectangular array termed as a bivariate or correlation table. It is often used by statisticians as a preliminary centre to show whether two series are significantly correlated.

If the probabilities of the two variables have been given rather than their frequencies, the distribution is then a bivariate probability distribution. The distributions obtained by adding along each row or column of a bivariate distribution table have been-termed as marginal distribution.

Bivariate exponential distribution: It is any bivariate distribution that is having exponential marginal distributions. Some have been special cases of bivariate gamma distribution which are obtained by setting the index parameters to unity. Others include Gumbel's bivariate exponential distribution with probability density function

$$h(x, y) = \{(1+ax)(1+ay)-a\}\, e^{-x-y-axy}$$
$$x \geq 0,\ y \geq 0, a \geq 0,$$

and the bivariate case of the Marshall-Oklin distribution.

Bivariate-F-distribution. It is a bivariate generalisation of the F-distribution which arises naturally in the simultaneous testing of linear hypotheses in the analysis of variance.

Bivariate gamma distribution: Refers to any bivariate distribution having gamma marginal distribution. Examples of such distributions are cited by Wicksell, McKay and Cherian.

Bivaraite logarithmic distribution: It is discrete bivariate distribution which was introduced by Khatri. It is having joint probability mass function.

$$\Pr(x, y) =$$

$$\frac{(x+y-1)!}{x!y!} \cdot \frac{p_1{}^x p_2{}^y}{-\log(1-p_1-p_2)}$$

where x, $y = \geq 1$, $0<p_1$, $p_2<1$, and $p_1-p_2<1$, and arises as the limit of the (0, 0)-truncated bivariate negative, binomial distribution. Alternatively it is termed bivariate logarithmic series distribution.

Bivariate negative binomial distribution: It is discrete bivariate distribution having joint probability mass function.

$$\Pr(x,y) = \frac{(x+y-k-1)}{x!y!k} p_1^x p_2^y (1-p_1\ p_2)^k$$

where x, $y = 0, 1...$, $k>0$, $0<p_1$, $p_2<1$, and $p_1+p_2<1$. The conditional distributions $\Pr(x|y)$ and $\Pr(y|x)$ have been negative binomial.

Guldberg was first of all introduced this distribution, and Lundberg and first used it in connection with problems of accident proneness.

Bivariate normal distribution: Two variates x_1 and x_2 having means m_1 and m_2 and variances σ_1^2 and σ_2^2 get distributed in the bivariate Normal form if their distribution function has been given by

$$dF = \frac{1}{2\pi\sigma_1\sigma_2(1-p^2)^{1/2}} \exp$$

$$\left[-\frac{2}{2(1-p^2)}\left\{\left(\frac{x_1-m_1}{\sigma_2}\right)^2\right]\right.$$

$$-\frac{2p(x_1-m_1)(x_2-m_2)}{\sigma_1\sigma_2}+$$

$$\left.\left(\frac{x_2-m_2}{\sigma_2}\right\}\right)dx_1 dx_2,$$

where p denotes a parameter not greater than unity in absolute value. For any fixed x_1 (or x_2) the variate x_2 (or x_1) is Normally distributed. The parameter p has been (or ought to be) called the correlation parameter.

Bivariate pascal distribution: Suppose an experiment of observing individuals is considered one after another for two characters A_1 and A_2 until exactly k individuals having both the characters are observed. If x_1 denote the number of individuals possessing character A_1 observed in the experiment, i = 1, 2, the x_1 x_2) has the bivariate Pascal distribution. This distribution has been a special case of the bivariate negative binomial distribution.

Bivariate poisson distribution: It is discrete bivariate distribution having frequency function of the form $f(x, y; \theta)$

$$= \exp(\theta-\lambda_1-\theta_2)$$

$$\sum_{k=0}^{\min(x,y)} \frac{\theta^k(\lambda_1-\theta)^{x-k}(\lambda_2-\theta)^{y-k}}{k!(x-k)!(y-k)!}$$

$x = 0, 1, 2, ...; y = 0, 1, 2, ...$

where $0 \leq \theta \leq \min(\lambda_1, \lambda_2)$. The marginal distributions of the random variables X and Y have been Poisson with parameters λ_1 and λ_2 respectively.

Bivariate 'student' distribution: It is a bivariate generalisation of 'Student's t-distribution' which was introduced by Pearson for graduation purposes. The distribution is having the probability density function

$$h(x, y)=\{2\pi\sigma_1\sigma_2 \sqrt{(1-p^2)^{-1}}\ \frac{f}{f-2}$$

where $-\infty<x,\ y<\infty, j>0$. The random variables X/σ_1 and $Y/\sigma2$ are having marginal t-distributions with degrees of freedom, and X and Y are having product moment correlation coefficient p.

Bivariate type II distribution: It is a continuous bivariate distribution having probability density function

$$h(x, y)=\{2\pi\sigma_1\sigma_2 \sqrt{(1-p^2)^{-1}}\ \frac{n+1}{n+2}$$

$$\times\left[\ 1-\frac{1}{2(n+2)(1-p^2)}\right.$$

$$\times\left\{\frac{x}{\sigma_1^2}-\frac{2pxy}{p_1p_2}+\frac{y}{\sigma_2^2}\ \right\}$$

where $-c\sigma_1 < x < c\sigma_{1,} -\sigma 2 < y < c\sigma_2$, $c^2=2(n+2)$ and $n>0$ introduced by Pearson. The marginal distributions of both X and Y have been type II distributions.

Bivariate uniform distribution: It is any bivariate probability distribution that is having uniform marginal distributions. Examples have been the contingency-type uniform distribution, otherwise known as Plackett's uniform distribution, and Morgenstern's uniform distribution.

Blind test: A technique for evaluating consumer's reactions, to products, especially food, in which a group of people are given unidentified samples and asked for their comments.

Block: Refers to a subset of the items under investigation in statistical experiment. It could be:

The barley in a particular part of a field.

The householders in one street of a town.

The weekly earnings of one particular employee of a factory.

The variability within a block would be expected to be less than that within the whole population being investigated.

For example, a block may comprise a group of contiguous plots of land, all the animals in a litter, the results obtained by a single operator, or meteorological observations on a series of .days at a given place. The general purpose of dividing all the material in an experiment into blocks (or of regarding it as so divided by' the circumstances of the case) has been to isolate sources of heterogeneity; the items in a block being so far as possible homogeneous and uncontrolled variation in the experimental material being measured by comparisons between blocks.

The variation in the experimental observation has been usually divided (by variance-analysis) into effects due to differences between blocks and effects, due to variation within blocks; these being known as interblock and intrablock effects (variances, etc.) respectively. The expressions 'interblock' and 'intrablock information' occur both in the general sense and in the specialised sense of information. Thus for estimates of certain treatment comparisons the 'information' may be the reciprocal of the most efficient estimator if it exists.

Block diagram: It is made up of vertically placed rectangles which are situated adjacent to each other on a common base line. Where the characteristic to be depicted has been quantitative the height of the rectangles is generally considered to be proportional to this quantitative variable. When this kind of diagram is used to portray a frequency distribution it takes the .name of histogram.

Boole's inequality: An inequality which was developed by Boole in 1854 to provide the limits to the frequencies in certain logically defined classes in terms of the frequencies in other classes. It finds application in probability theory., One simple form states that if $A_1, A_2, \ldots, A_k$ have been compatible eve-its the probability that at least one occurs will not be greater than the sum of the probabilities that each occurs independently of the occurrence of the others:

$$1\text{-}Pr(\text{not-}A_1, \text{not-}A_2. \ldots.. \text{not-}A_k) \leq$$

$$Pr(A_1)+Pr(A_1)+Pr(A_2)+ \ldots +Pr(A_k)$$

Borel-Tanner distribution: In queuing theory it refers to the distribution, under certain conditions, of the number of customers served before the queue gets vanished for the first time.

Bose distribution: Refers to the distribution of the variance ratio when the variates get correlated. It was given by R. C. Bose.

Bose-Einstein statistics: In statistical mechanics one possible basic assumption concerning states and energy levels in equivalent to supposing that r distinguishable particles get distributed among n cells ($r < n$) in such a way that each of the n^r arrangements has been equally probable. This results in the Maxwell-Boltzmann statistics. If the particles have been indistinguish. able there have been $\binom{n+r-1}{r}$ distinguishable arrangements and if these are taken as equally probable there result Bose-Einstein statistics. As a particular case, if not more than one particle may appear in any cell there are $\binom{n}{r}$ equally probable arrangements; and these are forming the basis of Fermi-Diras statistics.

Box three boxes game: A game in which there are three boxes marked 1, 2, and 3. For a given play of the game, player A removes the bottom of one of the boxes, but player B does not know which one it is. Player B then puts an amount of money equal to the number marked on the box in each of two of the three boxes. He loses the money put in the box with no bottom and wins the money put in the others. This is a zero sum game with imperfect informations. The payoff matrix does not have a saddle point and the solutions are mixed strategies.

The solutions are (0, ½,½) for A and $\left(\frac{3}{5},\frac{2}{5},0\right)$ for B, meaning that A removes the bottoms of boxes 2 and 3, each with probability ½, player B puts money into, boxes 1 and 2, or 1 and 3, with respective probabilities $\frac{3}{5}$ and $\frac{2}{5}$ (never in and 3).

The value of this game is 1 (with B the maximizing player.)

Box-and-whisker plot: A diagram which is used to display some important features of the distribution of experimental data . It consists of two lines,, (whiskers) drawn from points representing the extreme values of the lower and upper quartiles, and a box drawn between the quartiles which include a line for the median. A box-and-whisker plot is also termed as a boxplot.

The boxplot is drawn as in fig. The box, drawn between the lower and upper quartiles, contains 50% of the date. The two Wiskers, from the quartiles to the least or greatest values, each contain 25% of the data.

Box's test: It is an approximate test which is used for the equality of the variances of a number of Populations. The important feature of the test has been its robustness under departures from Normality in the Populations.

Box-Jenkins model: It is a model which is used for predicting non-stationary time series with or without seasonal variations. It has been found to depend on autoregression of the mth difference of the series, having a moving average residual.

Branching process: It is a stochastic process which describes the growth of a Population in which the individual members may have offspring, the lines of descent 'branching-out' as new members have been born. Sometimes it is referred to as a chain reaction process. If the life times distribution of the individual members, i.e., the waiting time in a particular state, follows the negative exponential form the (continuous time) process called a branching Markov process. For a genera(form of life-time distribution the term age dependent branching process has been employed used.

Branching renewal process: It is a renewal process in which each of a series of primary events generates a subsidiary series of events. The complete process involves the superposition of the events in the primary and subsidiary processes.

Bravais correlation coefficient: It is an obsolete synonym for the correlation parameter in bivariate Normal variation; and, by extension, to the, productmoment correlation coefficient which estimates it.

Breakdown bound: Refers to one of a member of similar measures which are used to compare the robustness of estimators in the presence of extreme outliers. The bound provides an indication of the largest proportion that an estimator may be expected to tolerate.

Break even analysis: A technique to determine the rate of sales in quantity so that the rate of money sales can be equated to the rate of costs; usually expressed in terms of annual rates. There are two approaches: one, the linear, in which it is assumed that the selling price cif the product remains constant, differences in the rate of sales being due to other factors; the other, the curvilinear, in which it is assumed that differences in the rate of sales are due to differences in the selling price.

Linear approach. Figure 1 demonstrates this. The cost graph, T, is a straight line intersecting the vertical axis at a point, F, which is the measure of the fixed annual costs (F8), and S is the sales graph, which-is also a straight line since the price is held constant for all rates of sales. The point of intersection of the two graphs is the break-even point, Q, on the horizontal axis being the break-even quantity, S, on the vertical axis, the break-even money sales. Figure 1 shows how the problem is usually depicted in the literature.

Curvilinear approach: Figure 2 provides 6 diagrammatic demonstration of the curvilinear analysis. The graph of total cost, T, remains a straight line, as in 1, but the sales graph, S, becomes a curve, concave downwards as it rises to the right. Here, there are two break-even points, namely B_1 for a relatively high price, and B_2 for a relatively low price. Commercial undertakings are less interested in the break-even points in this kind of analysis than in the optimum price which the diagram indicates. The optimum price is the price that yields the maximum net profit, the measure of which is represented by the vertical distance between the cost graph and the sales curve where they are the furthest apart, i.e., where they are parallel.

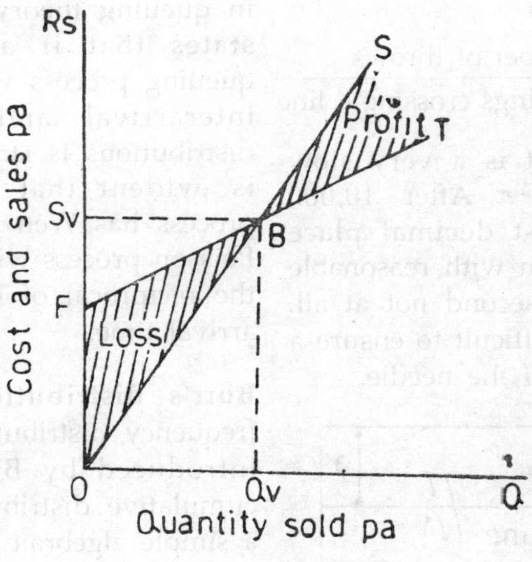

Fig. 1. Quantity sold per an.

Break-even chart: A chart on which two curves are plotted the expected sales revenues from the production of various quantities of a product and the costs (fixed and variable) of producing those quantities. The curves intersect at the break-even point.

Brownian motion process: An additive stochastic process in a real, variate x_1 defined at time t such that $x_t - x_s$ has been Normally distributed with zero mean and variances $\sigma^2 | r - s |$, where σ^2 has been a constant.

Butfon's needle: Refers to a statistically- based experiment for determining π. A piece of paper has been marked with parallel lines, distance a apart, and needle has been thrown onto the paper in a random manner. The probability that the needle crosses one of the lines has been put as follows:

$$\frac{2l}{a\pi}$$

where 1 is the length of the needle, and $1 < a$. Thus if the needle is dropped a large number of times, it becomes possible to estimate π as

$$\frac{2l}{a} \times \frac{\text{Total number of throws}}{\text{Number of landings crossing a line}}$$

This experiment is a very slow way of finding π. After 10,000 throws, the first decimal place should be known with reasonable confidence, the second not at all. It is also very difficult to ensure a random throw of the needle.

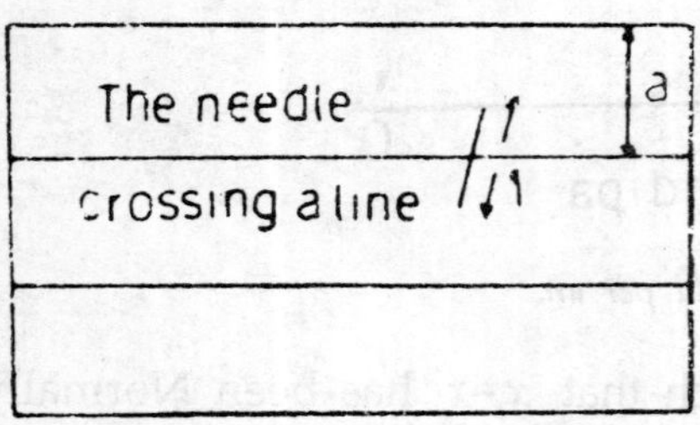

Buffon's needle

This experiment was originally intended as a method of calculating the stakes and winnings in the French gambling game france-carreau, rather than a method of estimating π.

Bulk sampling: Refers to the sampling of materials which have been available in bulk form. That is to say, it has been the population which is in bulk; the term does not mean the drawing of a sample in bulk. Examples of such sampling would be involving the sampling of a shipment of coal for ashcontent, or of tobacco for moisture content.

Burke's theorem: It is a theorem in queuing theory. This theorem states that if a single-server queuing process with exponential interarrival and service-time distributions is stationary, then it is evident that the departure process has Wen a homogeneous Poisson process with rate equal to the reciprocal of the mean inter-arrival time.

Burr's distribution: It is a frequency distribution which was introduced by Burr where the cumulative distribution is having a simple algebraic form

$$F(x) = 1-1/(1+n^c)^k \quad n \geq 0; \; c, k \geq 1.$$

It has been one of a class wherein the distribution function, not the-frequency function, has been defined in analytical terms.

Busy period: In queuing theory it refers to a closed time interval such that all servers in a queuing system get occupied.

C

Campbell's theorem: Refers to an evaluation of the asymptotic distribution of the sum of the effects of random impulses which are acting with given intensity on a damped system.

The impulses are considered to occur at times according to Poisson process with parameter λ. Each impulse is having given intensity α and is having an effect $\alpha\psi(t)$ after time t has, passed. If θ (t) be the sum of the effects of all impulses occurring prior to time t, then for $t\to\infty$ the mean and variance of θ (t) will be respectively.

$$\alpha \int_0^\infty \psi(t)dt,\ \lambda\alpha^2 \int_0^\infty \{\psi(t)^2 dt$$

Canonical matrix: In general, it refers to any matrix which reduces to a canonical form; e.g. , in the analysis of canonical correlations. It has been sometimes assigned specialised meaning.

Cantor-type distribution: It is a thoroughly pathological type of probability distribution for which there' exists distribution function F and is continuous but for which there exists no density function.

Capon test: It is a normal scores test for a difference between the scale parameters of two otherwise identical populations. The ranks R_i, i = 1, 2, ... m of the members of a random sample X_1 X_2,... X_m. from one of the populations, after getting combined with a random sample Y_1, Y_2, ... Y_n from the other, get transformed to scores which may be defined as the expectations of the squares of the order statistics of rank R_i in a random sample of $m+n$ realisations from a 'unit Normal distribution; the sum of these scores has been the test statistic.

Capture/release sampling: It is a method of sampling which has been found to be specially suited to the estimation of the size of total populations of wild animals. It is also called capture/recapture sampling. The method was mainly introduced and used by Lincoln. This method involves capturing, marking and releasing a random sample, say, of animals of a particular kind. Subsequently, a further random sample is, consid-ered and the proportion of marked animals in this sample will form the basis of estimates of total populations.

Cardinal number: It is a number which is denoting quantity but not order in a group. Thus 1, 2, 3. ... have been cardinal numbers. The word cardinal has been generally used in contrast to ordinal which expresses position in a sequence, e.g. first, second, third etc.

Cardioid distribution: It is a circular distribution for which the density function is having a cardioid curve as its polar representation. It was given by Jeffreys, and is having the density function

$$f(\theta) = (2p)^{-1}\,[1+2_{pcos}(\theta-\theta_0)],$$
$$0<\theta \leq 2\pi,\ |p| < ½$$

For $p = 0$, it gets reduced to the circular uniform distribution

Carleman's criterion: There are certain conditions under which it is possible for two different distributions to have the same set of moments. Criteria therefore become desirable to decide when a set of moments will be able to determine a distribution uniquely. One such, given by Carleman (1925), states that a set of moments μ_1, μ_2..., not necessarily about the mean, will be able to determine a distribution uniquely if:

$$\sum_{j=0(\mu 2j)}^{\infty} \frac{1}{1/2j} \text{ for distributions}$$

ranging from — ∞ to ∞)

or

$$\sum_{j=0(\mu j)}^{\infty} \frac{1}{1/2j} \text{ (for distributions}$$

ranging from 0 to ∞)

diverges. Cramer and Wold (1936) generalised this to the multi-variate case.

Carli's index: It refers to a simple index number of prices. It was proposed by Carli in 1764. Let the prices of a set of commodities in the base and given period are respectively p_0, p_0', p'', ...1 and p_n p_n', p_n'', ..., Then, Carli's index number will be given by

$$I_{on} = \frac{1}{k}\sum\left(\frac{p_n}{p_o}\right)$$

where k denotes the number of commodities. The summation extends over all commoditions. It is therefore an unweighted index of price-relatives.

Carrier variable: It is a name assigned to some numerical quantity for which the different values furnish the levels of the corresponding factor in a factorial experiment.

Cartogram: It is a device which is used for displaying statistical information of a descriptive nature by using a symbol on a map. The symbolism may get various forms in accordance to taste, *e.g.,* dots or circles of varying density, or shading in black and white, or use of a full range of colours. The various forms of cartogram have been found to be especially convenient for portraying data according to geographical distribution.

Categorical distribution: A distribution is considered to be categ-orical if the data could be sorted into categories in accordance, to some qualitative description rather than by a numerical variable.

Categorical scale or nomiral scale: It refers to a scale which sorts variables in accordance to category. For example, the cars in a park can be sorted according to make, the fish in a pond according to type, etc. For a categorical scale t becomes essential that the categories be exhaustive, exclusive and clearly defined. Hence every variable should belong to one and only one category, and there should be no doubt as to which one.

There have been two types of categories, a priori and constructed. A priori categories already exist, for example, a person's sex or year of birth; constructed categories have been decided upon by research workers. A bad choice of categories outcome of an investigation. Categories may get indexed by numbers which do not correspond to any true order, such ad the numbers on footballers's shirts.

Category: It is a homogeneous class or group of a population of objects or measurements. It is possible to style the category after one of the finite characteristics of the population or according to the limits of measurement for which observations are to be allocated to that category of frequency group. For example, it is possible to categorise people according to Sex (Male or Female) Or Age next birthday (1 -5 years; 6-10 15-15; 16-20; and so on)

Cauchy distribution: It is the name generally assigned to denote the continuous distribution

$$dF = \frac{dx}{\pi (l + x^2)}, -\infty \leq x \leq \infty$$

or some simple transform to another origin or scale. The distribution is having no finite moments other, perhaps, than the arithmetic mean, which itself may be regarded to exist only by convention.

It has been a particular case of Student's distribution having one degree of freedom.

Causal distribution: It is an alternative designation for deterministic distribution. As it is neither causal nor a distribution except in a degenerate way, it should be avoided.

Causal relationship: Statistical analysis can go a long way towards establishing the existence of a causal relationship between two measurable factors. It cannot establish the nature of any causal relationship, nor can it be used for proving that any two factors are not causally related.

So far as the nature is concerned, statistical analysis may Show that factors A and B are causally related; It cannot show that A causes B or that B causes A, or indeed, that either one causes the other, since the relationship shown to exist between A and B may be the effect of factor C, i.e., it may be that A and 13 are joint effects of C.

As to factors that, on the evidence of the statistics, seem not to be causally related, B and C may be joint causes of A, but C may behave so erratically relative to B that the statistics of A and B fail to reveal any causal relationship. Table contains a hypothetical example of a problem in three variables, in which A is dependent on 133 and C, and in which the statistics for A are built up from the equation A = 4B + 10C.

Table Causal relationship with three variables.

A	*B*	C
158	2	15
148	5	12
120	10	8
134	11	9
132	8	10

A comparison by inspection of the two series representing A and B, or for that matter, A and C, suggests little causal relationship between the two factors. Yet if we could compare A with B and C together, we would find the statistical relationship to be 100%. There is a statistical measure of the relationship between two or more statistical series: it is called the coefficient of correlation.

An understanding of the logic of causal relationships is of use in statistical research in all fields.

Causation: It is the production of an effect possible to prove. Causation statistically, although very strongly suggested.

A common misunderstanding is concerned with the relationship between correlation and causation. If there happens to be a strong correlation between the variables measuring the occurrence of two events A and B, it may be that

A causes B, A ⇒ B or B causes A, B ⇒ A

or they both cause each other, A⟸⟹B

but it might also be that a third event C causes both A and B,

$$C \begin{matrix} \nearrow A \\ \searrow B \end{matrix}$$

or that a more complex set of interactor's is going on. It may also be that the correlation observed has been merely due to chance fluctuations in the variables involved and that there has been no causal reason for it at all. The statement 'Correlation does not imply causation' explains this situation.

The fact that there is a strong (negative) correlation between the mean number of cars per family and the mean number of children per family for the various countries of the world does not mean that owning a car does not allow you having children, or that having children stops you having a car. The situation has been more complicated when the affluence of the country has been involved; in affluent countries, birth rates have been low and car ownership is common.

Affluence ↗ Low birth rate
↘ High car ownershi

(It could, however, be argued that affluence and low birth rates both cause each other).

Cause variable: If relation such as $y = f(x)$ has been interpreted in a causal sense, *e.g., y is* has been considered as caused by x, the latter is sometimes called a cause variable and the former an effect variable. The cause variable is also termed as an explanatory variable.

Cell frequency: If the classification of a frequency distribution is done into categories, univariate or multivari-ate, the subcategories are sometimes termed as cells; the frequency, with which observations come into a particular cell is termed as the cell frequency.

Censoring: We consider a sample to be censored when certain values are not known (or deliberately ignored) although their existence is known. The term is generally used in those cases where of a number n of values existing, only the k smallest or k largest are observed or k might be expressed as a proportion *of n.* This is termed as Type II censoring; Type I is when the sample gets censored by, reference to a fixed variate value. Whereas Type I censoring can be 'right' or 'left' censored it is more usual for Type II censoring to be only On the right of the variable. The ignoring of the existence of additional values or where such cannot exist is termed as truncation and refers to the population.

Census: The term, used for the

complete enumeration of a population or groups at a point in time with respect to well-defined characteristics for example, Population, Production, Traffic on particular roads. In some cases the term is associated with the data collected rather than the extent of the collection so that the term Sample Census possesses a distinct meaning.

The partial enumeration which gets resulted from a failure to cover the whole population, as distinct from a designed sample enquiry, could be referred to as an 'incomplete census'.

In general, consus refers to an official periodic count of a population, including such information as sex, age, occupation: In India census is held every ten years. Heads of house holds are required to provide information about all those living under their roots. The questions asked cover name, age family status, education and employment.

Centile: It is an abbreviated form of 'percentile'. It is not in general use, but frequently used in the statistical literature of psychological and educational testing.

Central confidence interval: We consider a confidence interval for a parameter θ with lower and upper limits t_1 and t_2 to be central when.

$$\Pr\{(\theta-t_1)<0\}=\Pr\{(t_2-\theta)<0\}.$$

The values t_1, and t_2 are then, in a sense, symmetrically placed with respect to θ.

Central factorial moments: In a strict sense, this expression means the factorial moments about the centre or some central value, Actually the term is sometimes applicable to the factorial moments which are calculated about some point near to a central value as distinct from the ends of the range.

Central limit theorem: A rule assuring that the sampling distribution of the mean approaches normal as the sample size increases, regardless of the shape of the population distribution from which the sample is selected.

This theorem states, taken if samples of size n are taken from a parent population with mean μ and standard deviation then the distribution of their means will be approximately normal with

$$\text{Mean} = \mu$$

$$\textit{Standard deviation} = \frac{\sigma}{\sqrt{n}}$$

As the sample size n increases, this distribution approaches the normal distribution with increasing accuracy (see sampling distribution). Hence in the limit, as n-∞, the distribution of the sample means → Normal, mean μ, standard deviation $\sigma/\sqrt{n}$.

If the parent population has

been itself normal, the distribution of the sample means will be normal, whatever the sample size; If the parent population has been of finite size two possibilities arise:

(a) If the sampling has been done with replacement, the theorem stands as stated;

(b) If there has been no replacement, the standard deviation of the sample means would be

$$\frac{\sigma}{\sqrt{n}}\sqrt{\frac{N-n}{N-1}}$$

The factor $\frac{\sigma}{\sqrt{n}}\sqrt{\frac{N-n}{N-1}}$ has been termed as the finite population correction.

The central limit theorem forms the basis for much of sampling theory; it can be summed up, as follows. If n is not small, the sampling distri-bution of the means is approxi-mately normal has mean = μ (the parent mean), and has standard deviation = $\sigma/\sqrt{n}$.(where σ denotes the parent standard deviation).

This theorem, although due to Laplace, was first proved rigorously by Liapounov.

Central tendency: Refers to the tendency of quantitative data to cluster around some variate value. The position of the central value is generally determined by one of the measures of location like the mean, median or mode. The closeness with which values cluster round the central value is measured by one of the measures of dispersion like the mean deviation or standard deviation.

Centre (of a range): In a specialised sense the phrase centre of a sample' sometimes denotes the variate value which is midway between the-two extreme variate val ıes; correspondingly, centre' is sometimes used to refer to the mid-point of the range of a distribution.

Certainty: Refers to the condition of an event whose probability of occurrence has been equal to 1. It has been the decision environ-ment in which only one state of nature exists.

Chain: It refers to a sequence of terms in such a way that each term is depending in some defined way upon the previous term or terms in the series; for example, the chain-relative used in the calculation of index num-bers upon the chain-base method.

The term chain has been also used in connection wit stochastic processes where the value at one point has been determined by Values at previous points apart from a random element; or more exactly, the probability distri-bution at any point, conditional on certain previous values, has been otherwise independent of past history, The most common case has been the Markov chain.

Chain index: Refers to an, index number in which the value at any given period gets related to a base in the previous period, as distinct from one which gets related to a fixed base. The comparison of non-adjacent periods is generally made by multiplying consecutive values of the index numbers, which, as *it* were, form a chain from one period to another. For example, if the value of the index for period 2 based on period 1 has been I_{12} and that for period 1 on period 0 has been 101 the chain index for period 2 based on period 0 is $I_{01} \times I_{12}$ (divided by 100 if the index numbers are based on 100 as the standard).

In practice chain index numbers are generally formed from weighted average of link-relatives, namely the values of magnitudes for a given period divided by the corresponding values in the previous period.

Champernowne distributions: Refers to a three-parameter system of distributions which was suggested by Champernowne in connection with the distribution of incomes:

$$F(x) = 1 - (\cos\theta)\tan^{-1}$$

$$\left[\sin\theta \Big/ \left\{\cos\theta + \left(\frac{x_1}{m}\right)^{\alpha}\right\}\right]$$

where m denotes the median of x; θ and *a* have been constants.

Chance: 1. Refers to the unknown and unpredictable element that causes an event to have one outcome rather than another. 2. Refers to that which is described by probability.

Chance variation: Refers to variation in statistical observations due to the action of random, as distinct from systematic, factors.

Changeover trial: Refers to an alternative name for an experiment or trial which is using a crossover design.

Charlier's check: It is an accuracy check when can the used while we calculate the mean and standard deviation. It employs the fact that for a set of numbers x_1, x_2,.... x_n with frequencies f_1, f_2,.... f_n

$$\sum_{i=1}^{n} f_i(x_1+1)^2 = \sum_{i=1}^{n} f_i x^2 i + 2$$

$$\sum_{i=1}^{n} f_1 x_1 + \sum_{i=1}^{n} f_i$$

Chebyshev's inequality: The theorem states that: If a probability distribution is having mean μ and standard deviation σ, then the probability that the value of a random variable with that distribution differs from μ by more than kσ is less than $1/k^2$, i.e.,

Chebyshev's ineorem: No matter what the shape of a distribution, at least 75 per cent of the values in the population will fall within 2 standard deviations of the mean, and at least 89 per cent will fall within 3 standard deviations.

Chi distribution: Refers to the distribution of the positive square root of the statistic called the χ^2 or chi-squared.

Chi-squared statistic: In a strict sense this expression should refer to a statistic which is distributed a's chi-squared (X^2) namely as the sum of squares of independent standard Normal variates. For historical reasons however, it more usually relates to a statistic of a particular kind which is, as a general rule, distributed more or less approximately in the χ^2 form. If a set of n values gets distributed over *k* classes such that the observed frequency in the *j*th class-has been *nj* and the theoretical (expected) frequency in that class has been v_j, the Statistic.

$$\sum_{i=1}^{k} \frac{(n_j - v_j)^2}{v_j}$$

has been called the χ_2 Statistic, or simply the value of χ_3, for the data. It has been widely used to test agreement between observation, as-represented by the *nj* and hypothesis as represented by the *vj*

Chi-spuared (or χ^2) test: It is a test of how well observed data fit a theoretical distribution, The data have been collected into groups, and the number of observations in each group denoted by 0 (observed). The expected number in each group E has been calculated from the theoretical distribution. The value of χ^2 has been then worked out using

$$\chi^2 = \sum_{All\ groups} \frac{(O-E)^2}{E}$$

A small value of χ^2 implies the data fit the distribution well, a large value badly.

Chi-Squared test: Occur in many ways, the most prominent being:

(i) An overall goodness-of-fit comparison of observed with hypothetical frequencies falling into specified classes;

(ii) Comparison of an observed with a hypothetical variance in normal samples;

(iii) Combination of probabilities from a number of tests of significance.

Chi-Squared test for variance.. It is a test of the null hypothesis, that a particular sample is drawn from a parent population having a given variance. This test can only be applied if the parent population has been normal. It is dependent upon the fact that the statistic

$$\frac{n\sigma^2}{\sigma^2}$$

is having a χ^2 distribution with

n—1 degrees of freedom, where n has been the sample size, a has been the parent standard deviation estimated from the sample and σ has been the given parent standard deviation.

Chi-statistic: It is equal to the square root of the more familiar chi-squared statistic.

Cigarette card distribution: If there have been k different types of cigarette cards being mixed equally frequently and the types are randomly distributed, the expected distribution of the k types in a collection of n cards ($n > k$) has been

$$\frac{k!}{(k-r)!r!}\ \frac{\Delta r(0^n)}{k^n}$$

where the second term is using the leading differences of powers of natural numbers.

Circular chart: It is a method of diagrammatic representation whereby the components of a single total could be shown as sectors of a circle. The angles of the sectors will be proportional to the components of the total. Additional Visual aid could be obtained with coloured shading or cross-hatching. Also termed as a pie chart.

Circular distribution: It is a frequency distribution of a variate which ranges from 0 to 2π, so that the frequency may be considered as distributed round the circumference of a circle. The term has been used especially of phenomena which have a period of 2π = (by a suitable change of scale if necessary) so that the probability density at any point a would be the same as that at any point $\alpha+2\pi r$ for integral values of r. It has been generilly expressed in terms of an angle θ. The analogue of the Normal distribution in this class, the so-called circular Normal distribution, would be given by

$f(\theta) = e^k\cos(\theta-\theta_o)/2\pi I_o(k),\ 0\leq\theta<2\pi,$

where I_o, denotes a Bessel function of the first kind of imaginary argument. It was derived by von Mises in 1918. I ispossible to derive the expressions analogous to the standard deviation, variance, Pearson measure of skewness and kurtosis for the general circular distribution. For example Mardia (1972) defines, for a set of values $\theta_1, \theta_2,....\theta_n$ the mean direction θ

as $\sum_{j=1}^{n} 0_j/n$ and the circular variance as

$$S_o = 1-\frac{1}{n}\sum_{j=1}^{n}\cos(\theta_j-\theta_0)$$

It could also be defined as $\Sigma(\theta_j-\theta_o)2/n$ to which S_0 reduces for small θ_j—θ_0.

The circular uniform distribution would be the simple form

$$f(\theta) = 1/(2\pi),\ 0 \leq \theta \leq 2\pi$$

Mardia also defines a circular triangular distribution, in, symmetric form about $\theta = \pi$, as

$$f(\theta) = \frac{1}{8\pi}\ \{4-\pi^2 p+2\pi p/\pi-\theta/\},$$

$$0\leq\theta\leq 2\pi,$$

and in asymmetric form as

$$f(\theta) = \frac{1}{2\pi}\{1+p(\theta-\pi)\},\ 0\leq\theta<2\pi,$$

$$0<p<1/\pi.$$

Circular histogram: It refers to a form of diagrammatic representation for grouped angular data. In contrast to the linear histogram where the rectangular blocks representing frequencies have been erected on a horizontal base, the circular, histogram is having the blocks attached to the, appropriate intervals on the circumference of a circle.

Circular test: In constructing index numbers a decision has to be made-as to the period upon which to base the index. If an index for period A based upon period B has been *IBA* and for period B based upon period C has been *ICB*, the circular test, derived by Irving Fisher, needs that the index for period A based upon period *C, i.e.*, IBA, should remain the same as if it were compounded of two stages, the calculation of A on B and that of B cn C. That is to say we should have

$$I_{CA} = I_{CB}I_{BA}.$$

A similar argument is applicable to comparisons between A similar argument is applicable so have been found to places. Few index numbers in current use have been found to satisfy this test.

Class: Refers to a collection or divisiorr of people, things, or values of la variable, sharing it common characteristic.

The world 'class' has some mild specialisation in the theory of frequency distributions. The total number of observations made upon a particular variate could be grouped into classes according to convenient *divisions* of the variate range in order to make subsequent analysis less laborious or for other reasons. A group so determined is termed as a class. The variate values which-determine the upper and lower limits of a class are termed as class boundaries; the interval between them has been the class interval; and the frequency falling into the class has been the class frequency.

Class boundaries: Refer to the boundaries for the classes when data has been grouped. For a continuous variable, the upper boundary of one class may be remain the same as the lower boundary' for the-next one. The reason for this has been that, in theory, if measured accurately enough, the variable will virtually always be one side or the other of the boundary. Hence the heights of adult men (in metres) could be grouped as

1.65—1.70 1.70—1.75
1.75—1.80 etc.

In practice, however, measurements have been always rounded, in some way or other, and so it often becomes convenient to make it clear which class a variable apparently on the boundary has been to be assigned to. In that case the classes, might be described as:

1.65—1.70—1.75 -etc.

or 1,65 $\leq$ <1.70   1.70 $\leq$ <1.75
1.75 $\leq$ <1.80

For a discrete variable, on the other hand class boundaries could not be shared.

Class frequency: The number of elements or individuals who fall in a particular class is called the class frequency. In the table below, the class frequencies are 108, 56, 55, 13 1.

	Age	*Frequency*
Class 1	**0-19**	108
Class 2	20-29	56
Class 3	30-39	55
Class 4	40 and over	131
Total		350

Classical probability: The number of outcomes favourable to the occurrence of an event divided by the total number of possible outcomes.

Classification: For the purpose of convenient analysis, observations may be grouped into classes, so that the classes together exhaustively cover the entire range of values. The process of classification *(i.e* , the determining of class intervals, and consequently of class frequencies) is an arbitrary process. Classes are humaniy determined, whereas attribute categories *(e.g.* sex) are usually self-evident.

In contrast with this general meaning of classification, the term classification statistic means a statistic used to determine from which population a sample originated, when it may have been drawn from one of a number of populations. In this context the term 'classification' assumes its more general nonstatistical meaning of discrimination between one population and another.

Class interval: The range of interval between the highest and lowest values allowed in a particular Class. For example, if we classify football terms by the number of points they obtained in the, Football League last year, we might have a frequency distribution like that given below:

Number of points	*Frequency*
Fewer than 25	5
25-34	24

35-44	25
45-54	20
55 and over	18
Total	92

Note that two of the class intervals are open—i.e., in the lowest class no lower limit is defined (although 0 could be assumed), and in the highest class no upper limit is defined. The other three classes are closed and the interval is 10 points.

Class limits: The smallest and large given class; there are stated limits and real limits.

Class mark: A term which is sometime used in elementary statistics, but obsolescent for advanced work, for denoting midvalue of the class interval.

The midpoint of a class—that is, its class mark—is calculated by averaging the lower and upper limits of that class.

$$\text{Class mark} = \frac{\text{Stated lower limits + Stated upper time}}{2}$$

When we are dealing with discrete variables and wide intervals, the midpoint of a class (its class mark) is calculated by using a slight modification of Equation (1) This enables us to work with more convenient values,

$$\text{Class mark} = \frac{\text{Stated lower limit+Stated lower limit of the next class}}{2}$$

Class symbol: In the theory of attributes, it refers to a letter which denotes membership or non-membership of a class, e.g., if A represents 'male' and α 'female', β represents living and β 'dead', AB would represent 'living males'. In Yule's notation the symbol (AB) would represent the number of living in the population under discussion.

Sometimes the symbol $\overline{B}$ has been used instead of β to denote not-B.

Classification statistic: In general, it refers to a statistic which is calculated from a sample for the purpose of assigning the Population from which the sample emanated to one of a number of classes. The term is some what synonymous with discriminant function.

Clisy: It is a term which was introduced by K. Pearson in connection with bivariate frequency arrays. For any fixed value of one variate x the distribution of y's is having a third meanmoment μ_3 (z): the way in which $\mu_3(x)$ varies with x, or the corresponding $\mu_s(y)$ and y, expresses the variation in clisy of the distribution.

A curve which shows values of skewness for the different frequency arrays of one variate is termed as clitic Curve. In plotting these curves Pearson apparently used the measure of skewness β_1

$(=\mu_3^2 / \mu_2^3)$ Kendall employed the term in a slightly different sense to denote the graph of the third moment of the arrays against the corresponding variate value.

If all arrays ate having the same skewness they are considered to be homoclitic; if not, hetero-clitic; Pearson also defined a hetero-clitic system, as nomic or anomic according as the skewness changes continuously or irregularly with the position of the array, but the terms are not into general use.

Closed question: A question in which the respondent is given a limited number of responses from which to choose.

Closed sequential scheme: In sequential analysis the sampling generally continue until either an acceptance or a rejection boundary gets reached. The sample size has been not fixed but so as to avoid having (although perhaps, only rarely) to draw large samples before reaching a boundary it may be desirable to fix an upper limit to the sample size. The scheme is then- termed as 'closed'. In the contrary case it is termed as 'open'.

Cluster: Refers to a group of contiguous elements of a statistical population. e.g., a group of people who are living in a single house, a consecutive run of observations in an ordered series, or a set of adjacent plots in one part of a field.

Cluster analysis: A generic term used to describe a large body of statistical procedures which are designed to group, or 'cluster', either elements or variables in a data set according to some well-defined characteristic. The objective of the analysis may be: (1) data reduction, i.e., reducing the mass of data to manageable proportion.' This is done: (a) by grouping similar elements together which will mean that it will be sufficient to consider the groups (of which there will only be a relatively small number) and not the elements (of which there may be a very -large number) in subsequent analyses; or (b) by forming clusters of variables (in particular attitudinal questions) so that summary scores for the respondents on the clusters of variables may be used instead of having to use all the answers to the individual questions. (2) To construct a typology or classification of elements or variables-this may be appropriate in using sets of symptoms to construct a classification of illnesses for example. (3) To form classes or clusters of elements with particular characteristics-for example, identify deviant groups in a population. A form of cluster analysis has been used to identify the characteristics of drunked drivers. Such as analysis could be used for prediction purposes.

Cluster (Point) process: Given a set of clusters which have been independently and identically distributed point processes, a cluster process consists of the superposition of all the clusters.

Cluster sampling: Refers to a method of sampling in which the members of the sample have been chosen from one or several groups (or clusters) rather than at random. From a statistical point of view it is less- satisfactory than simple random sampling, but it may well be more practical and/ or economical.

Cluster size: The number of elements in a cluster. In human populations, most clusters are of unequal size *e.g.,* schools, firms, towns. However, in some cases clusters of equal size can be found, e.g., the number of index cards in a file may be constant, or the number of dwelling units per block in a new housing development.

Cochran's Q: The statistic used for cochran's criterion representing a adaptation of the chi-squared statistic test for significant differences between three or more sets or sets or sample of matched of matched observations.

Cochran's Q-test: Refers to an alternative designation for Cochran's criterion in matched samples; not to get confused with the coefficient of association due to Yule.

Cochran's Test: It is a test which was due to Cochran (1941), for homogeneity of a set of independent estimates of variance it has been based in the ratio of the largest estimate of variance to the total of all the estimates.

Code: The term code denotes a number (a code number, or code) which is used to identify a category or class for categorized data. For example, in the frequency distribution below the numbers in the final column are the codes used to identify the, classes in analysing the data.

Calss	*Frequency*	*Code*
Under 15	78	1
15 and under 16	91	2
-16 and under 11	65	3
17 and under 18	83	4
18 and over	103	5

Coded data: Data which is translated from the form in which it was collected to one that is more convenient for analysis. In questionnaires, people are usually asked to reply by ticking appropriate boxes. When replies have been analysed, each box usually is having a number associated with it, and it is these numbers which are analysed rather than the tickets themselves.

When figures have been grouped, it has been often possible to simplify calculations of mean and standard deviation by a particular kind of coding of the data.

Coding: Once the initial editing has been completed, the next operation is to prepare the data for analysis. In most surveys this involves coding the responses, The procedure of coding involves three stages. The first stage is to decide how to categorize the responses; the second stage is to allocate numerical or symbolic values to the categories; the third stage is to allocate each individual response to the appropriate category. The principal purpose of coding is to provide a convenient way of summarizing the responses (or observations) in surveys so that the data can be analysed more easily, generally on a computer. The set of categories for a question or set of questions, together with the numerical or symbolic codes to be allocated to them, is called a coding frame.

Coding may in effect be done before the fieldwork is carried out by printing the categories of response and the codes on the schedule itself. In this case all the interviewer (or in a self-completion questionnaire, the respondent) needs to do is to ring the appropriate code. Otherwise the coding must be done at a later stage.

Coding error: Any errors which arise in the coding of the responses. The degree to which errors arise will depend oh the ability of the coders, on the suitability and completeness of the coding frame and on the degree to which the editing has ensured the completeness and accuracy of the responses.

Coding frame: The set of categories into which we classify the responses to a question or other stimulus, together with the numerical or symbolic codes we allocate to these categories, is called a coding frame. In many cases, the categories are self-evident and no serious problems arise.

Coding sheet: The form on to which the codes for the responses to a survey schedule are written in preparation for punching. In some cases, where all the questions have pre-coded response, the punching may be done directly from the questionnaire or schedule.

Coefficient: Refers to an index of measurement of characteristic. Examples have been the coefficient of alienation, and the coefficient of determination. 2. Refers to a numerical of constant factor in an algebraic expression. For example, 5 and 7 have been the coefficients of x^2 and x^3 respectively in the expression $5x^2+7x^3$.

Generally word coefficient has the same meaning as 'in mathematics, but occasionally It has, been used to denotes a dimensionless statistic, e.g., the moment-ratio β_2 as a coefficient

,of kurtosis or the coefficient of product-moment correlation. In this sense the word 'index' has been also used.

Coefficient of alienation: Refers to a measure of tne extent to which two random variables get unrelated. The coefficient of alienation is given as follows;

$$\sqrt{(1-r^2)}$$

where r represents the product-moment correlation coefficient

Coefficient of correlation: The square root of the coefficient of determination. Its sign indicates the direction of the relationship between two variables direct or inverse.

Coefficient of multiple determination, R^2: The fraction of the variation of the dependent variable that is explained by the regression. R^2 measures how well the multiple regression fits the data.

Coefficient of variation: A relative measure of dispersion, comparable across distributions, which expresses the standard deviation as a percentage of the mean.

Coefficient of variation=

$$\frac{\text{Standard deviation mean}}{\text{mean}}$$

Coefficient of variation is a standardized comparative measure of dispersion. It is equal to the standard deviation of a series of figures, divided by the arithmetic mean of the series. The standard deviation provides an absolute measure of the dispersion of a series in the same terms as the series itself—tons, yards, $S, number, etc. The coefficient provides a means of comparing the dispersions in two or more series.

Coherence: It is a property of criteria on which decisions under uncertainty have been based which ensures that decisions whose consequences have been manifestly undesirable are avoided.

Collapsed stratum method: Refers to a method for estimating the variance of the mean of a stratified sample where the sample has been based, upon two random selections from each of a number of strata of equal size and the number of non-respondents results in too few completed strata. Two or more strata get amalgamated (collapsed) to form one stratum.

Collectively exhaustive events: The list of events that represents all the possible outcomes of an experiment.

Combinatorial test: It is a test of significance in which the sampling distribution of the test statistic has been obtained by the algebra of combinatorial analysis.

Common factor: In factor

analysis the factor(s) have been classified in accordance to the, way in which they contribute to the variables under analysis. Any factor which appears in two or more variates is termed as a common factor. If the factor appears in all the variates it is termed as a general factor. If it is common to a group of variates it is termed as a group factor. A factor appearing in only one variate is considered to be specific.

Common factor space: In one geometrical representation, of a multivariate situation the variation has been considered as taking place in a space of which each factor represents a dimension. When there have been, say, m common factors and specific factors the whole factor space is one of $n = m+8$ dimensions. Of this the m dimensional sub-space has been the Common Factor Space.

Communicate: A description of the relationship between two states in a Markov chain where they are mutually accessible from each other, that is, if A affects B and B conversely affects A.

For example, in the entry branching process the state customers' affects 'accounts paid' (0.5) and 'accounts paid' also affects 'customers' (0.8). Irrespective of other consequent states of both, these two states affect each other, or communicate.

Comparative mortality figure: May be defined as the ratio of the standardised death rate to the crude death rate in a standard population. It may also be considered as an index number in the Laspeyres' form.

Comparative mortality index: It is a variant of the comparative mortality figure. It has been a weighted average death rate, where the weights have been the mean of the actual (current) population and the standard population both expressed in proportions on a common basis of absolute size. In this sense it is termed as an index number of the Marshall-Edgeworth Bowley form.

Compensating error: In general, it refers to any error which compensates for other errors. More specifically, a class of error having zero mean (unbiassed) and subject to the central limit effect, so that the occurrence of several errors will tend to cancel out and their effect get reduced as 'the errers cumulate. In this sense the term may not to be recommended.

Competition process: It is stochastic process which is a two-dimensional extension of the birth and death process.

Complementary events: Events which have been both exclusive and exhaustive. The events, 'The television is turned on' and 'The television is turned off,' have been considered as complementary. They cannot both happen at the same time (they are exclusive), and they cover all possibilities 'the are exhaustive).

Complete class (of tests): A class of tests has been complete if all every test outside the class there has been one within the class which has been uniformly better in the sense of a measurable decision function.

Complete coverage: A survey or census-can be described as complete if all of the elements in the population are included -in the enumeration, Such a survey or census involves complete coverage of the population.

Complete linkage/ farthest neighbour: A form of cluster analysis. As in single linkage analysis the two most similar elements are first linked together. However, in order that another element be added to a clurter, the additional, element must be sufficiently similar to all the elements already in the cluster, Thus the method will produce compact clusters. such that each of the elements in a cluster is closely related to (similar to) all the other-elements in the cluster. The method is sometimes referr-ed to as 'farthest neighbour analysis since whether an element is, included in a cluster depends on its similarity to the elements is included in the cluster, which is most different from it.

Complete system of equations: A term which is used mainly in econometrics for denoting the equations determining the behaviour of an economic system or part of such a system. The set or equations is considered to be complete when it includes all the determining equations governing the system, or a set from which all equations can be deduced. The point of emphasising complet-eness has been, that in some cases it has been possible to estimate parameters taking place in an incomplete set of equations, but the estimators may then be biassed, a fact of which attention was drawn by Haavelmo in 1943.

Completely randomised design: Refers to a very simple form of experimental design in which the treatments get allocated to the experimental units purely on a chance basis.

Complex demodulation: Refers to a variant of the technique of demodulation where the basic series or stochastic process has been multiplied by a complex non-random function.

Complex gaussian distribution: A univariate complex Gaussian distribution has been based upon a random variable $z = x + iy$ where-the real and imaginary parts have been each in $a\ p$

variate Gaussian (Normal) form. This concept, introduced by Goodman, N. R. (1963) has been important in the spectrum in analysis of multiple time series.

Complex table: It is a table which shows the classification of a set of data is accordance to more than two different features as distinct from the one or two features of the simple table. An example is that a human population might be tabulated in 114 complex table according to age, civilian status and sex. The complexity lies, not only in the manifold nature of the classification, but in the difficulty of printing the results in a convenient form. This method of tabulation is sometimes termed as multiple cross classification.

Complex unit: It is a statistical unit of record which is derived from a combination of two or more simple units. Examples are national income per capita or net-ton miles per-train-mile (which is equivalent to the average train load) or pounds-per-week-earned-per person.

Component analysis: It may be defined as a branch of multivariate analysis which represents, a *k*-dimensional variation as due to a number orthogonal components; fewer than k if possible, but if not, in such a way that a few components are able to account for as much of the variation as possible. The components sought in practice have been linear functions of the original variates.

Component bar chart: It is a bar chart which shows the component parts of the aggregate represented by the total length of the bar. These component parts have been shown as sections of the bar with lengths in proportion to their relative size. Visual presentation could be aided by devices of cross-shading or colours.

Components of variance: Refer to an alternative designation for Model II (or Second Kind) in analysis of variance.

Composed poisson distribution: A term which was introduced by Janossy et al. (1950) for describing the class of distributions now more commonly termed as Hermite distributions.

Composite hypothesis: A composite statistical hypothesis may be frequently defined as a statistical hypothesis which is not simple. It is not entirely satisfactory and in practice the expression usually refers to a hypothesis which has been 'composed' of a group of simple hypothesis. For example, the hypothesis that a frequency function has been normal with unspecified mean and variance is composite since there has been a double infinity of values of mean and variance which, when specified, would give rise to a simple hypothesis.

Composite index number: It is a rather vaguely defined term which relates to an index number for which the component series have been from groups which are different in nature. The definition has been some what arbitrary in practice because much depends upon the point of view of both the complier land user of the index.

Composite sampling scheme: Refers to a scheme in which different parts of the sample have been drawn by different methods; for example, a sample of a national population could be taken by some form of area sampling in rural districts and by a random or systematic method in urban districts.

Compound frequency distribution: The expression is found in three senses: (1) If Several-sets of individuals, each with a frequency-distribution, get mingled to form a single set, the frequency-distribution of the latter may be considered to be compounded of the separate distributions; *e.g.,* the distribution of heights of a population of human beings may get compounded of the distributions of heights of males and females.

(2) A frequency-distribution arising from the Convolution of variates may be considered to be compounded of the distributions of the individual variates. (3) If a distribution has been found to depend on a parameter θ which itself has a distribution, the distribution obtained by summing over θ has been said to be compound. This usage appears undesirable, because the word 'compound' then refers to the way in which the distribution was reached, not the properties of the distribution itself. So also, it appears preferable to refer to a distribution arrived at under (2) as 'convoluted'.

Compound hypergeometric distribution: It is a distribution which is produced by averaging the hypergeometric distribution for given x over all possible values of X according to a prior distribution. A number of well-known distributions may be regarded as special cases of the compound hypergeometric distribution. Examples are, the Polya, the binomial and the rectangular.

Compound poisson distribution: It is a name sometimes assigned to a distribution which gets resulted from a Poisson distribution of parameter λ where λ itself is having a distribution. If the distribution of λ's has been represented by dF (λ) ,the probability of observing the number k

$$\Pr(k) = \frac{1}{k!}\int_0^{\infty} e^{-\lambda}\lambda^k d^F(\lambda)$$

Computed *f*-Ratio: A statistic

used to test the significance of the regression as a whole.

Computed *t*: A statistic used for testing the significance of an individual explanatory variable.

Concomitance (Concomitanza): Refers to the relation between two variates in time; especially of the variation of two time-series in the same direction (positive concomitance) or in opposite directions (negative concomitance).

Concordance (Concordanza): Refers to a particular form of relationship *(connessione)* between quantitative variables, or qualitative variables when they are having comparable Modalities. If the variables have been such that positive values of one have been associated with negative values of another there has been said to be discordance *(discordanza),* that is to say, concordance tries to represent the sign as well as the intensity of the relationship.

Concordance, coefficient of: A coefficient which particularly measures the extent of concordance, and is concerned with the extent of agreement among a number of ranked sets irrespective of paired comparison, unlike the coefficient of agreement, which is based on numbers of pairs of identical rankings.

The formula is

$$C = \frac{12\sum d^2}{m^2 n(n^2 - 1)}$$

where d is the deviation of the sums of rankings from their common mean, $\left[\frac{m(n+1)}{2}\right]$ and where m is the number of rankings and n the number of objects.

Concurrent deviation: Suppose $(x_1, y_1), (x_2, y_2), \ldots, (x_n, y_n)$ be pairs of observations taken from some convenient origin such as the means of the *x's,* and *y's.* If *xi, yi* are having the same sign they are considered to exhibit concurrent deviation. A coefficient of correlation between x and y could be derived as the proportion of p of concurrent deviations to the total of n deviation.

Concurrent validity: If a scale is constructed as an indicator of behaviour or of some independently observable criterion, its validity may be measured by the degree to which it succeeds in measuring this criterion. When the alternative measurement is carried out at about the same time this is called concurrent validity.

Conditional probability distribution: If the variates x_1, $x_2, \ldots x_p$ have a joint probability distribution, the distribution subset of variates holding the others constant is called a conditional probability distribution.

Conditional regression: Refers to a regression which is estimated under certain conditions known a priori to apply, to some of the parameter concerned; for example' in estimating price and cross elasticities from time-series, data the income elasticities involved can sometimes be considered to be known from cross-section data. It might be better to find an alternative name, because of the intimate connection between regression and conditional distributions.

Conditional statistic: Refers to a Statistic whose distribution has been conditional, that is to say, depends upon some quantity which it held, constant; the quantity in question has been generally itself some function of the variates entering into the Statistic.

Conditional test: A test of significance has been sometimes difficult to apply as the distribution of the test statistic involves, unknown parameters of the parent population, This difficulty may sometimes be avoided by introducing restrictions on the sampling distribution, e.g., by regarding only samples which have the same mean as that of the observed sample. It is equivalent to making the inference in a sub-population of samples, which have a fixed mean. The distribution and the inference based on it have been then said to be conditional.

Confidence belt: Refers to the area between the upper and lower confidence limits.

Confidence coefficient: Refers to the measure of probability a associated with a confidence interval which expresses the probability of the truth of a statement that the interval will include the parameter value.

Confidence curves: A unified concept which was proposed by Birnbaum (1961) to include point, confidence limit and interval estimation.

Confidence interval: Refers to that interval within which a parameter of a parent population gets calculated (on the basis of sample data) to have a stated probability of lying. The larger the sample size *n*, the smaller will be the confidence interval; in other words, the more accurate will the estimate of the parent mean.

Confidence level: The probability that we associate with an interval estimate of a population parameter indicating how confident we are that the interval estimate will include the population parameter.

Confidence limits: Refer to the lower and upper boundaries of a confidence interval.

Confidence region: If several parameters have been under estimate it is possible to define regions in the parameter space such that there will be assigned confidence a that the parameters are lying within them. It is the generalisation of the confidence interval to the case of more than one parameter and the domain so determined is termed as the confidence region.

Configuration: A set of n observations on a variate could be represented as a vector in n-dimensional space. A number *k* of vectors in the space may be considered as having geometrical properties of inter-relationship independently of the coordinate system. In factor analysis, the arrangement of variate vectors among themselves, and without regard to any frame of reference, is termed as their configuration.

The expression 'configuration of a sample' is also found in a somewhat different sense. The n observations define a point in the *n*-dimensional space and any set of such points lying in a subspace, e.g., a hyperplane may be considered to have the configuration.

'Configurational sampling' has been sometimes used as a synonym for grid sampling.

Conformity (Conformita): This term refers to the agreement between experimental results and those expected in accordance to some theoretical scheme.

Confounding: It is a device whereby, in large factorial experiments the size of blocks gets limited by sacrificing some of the independent comparisons relating to the higher order interactions. These Particular interactions may get deemed unimportant or of little practical consequence from a policy point of view.

The totality of possible treatment combinations does not get replicated in each block but gets divided amongst blocks in such a way that the main contrasts can be made within blocks but the others are not distinguishable from contrasts between blocks, with which they are regarded to be confounded.

Conservative process: It is a stochastic process which governs the behaviour of a population which has constant total size but the members of which can assume independently one of a finite number of states, the variation having transfer from one state to another.

Conservative confidence interval: A confidence interval or region is considered as 'conservative' when the actual confidence coefficient gets exceeded the nominal or stated value.

Consistency checks: A part of the editing process in preparing survey data for analysis is to check whether the answers given to different questions are consistent with one another. One of the most common forms the checks take is to ask for the same information in two or more different ways. If, for example, at the beginning of the interview the respondent is asked for his date of birth, he might be asked at the end to give his age. The two answers should agree. Similarly if age, age at marriage and number of years of marriage are all obtained during the interview, then the number of years of marriage should be equal to age less age at marriage. The difficulty is that if the checking is done after the interview is completed and a discrepancy appears, there is no way of telling which of the answers is correct. However, the checks are valuable in providing an indication of the quality of the information collected.

Consistent estimator: It is an estimator which gets converged in probability, as the sample size increases, to the parameter of which it is an estimator. An example of an inconsistent estimator has been the sample mean as estimator of θ in the distribution

$$dF = \frac{dx}{\pi\{1+(x-\theta)^2}, -\infty \le x \le \infty$$

Consistent test: A test of a hypothesis has been consistent with respect to a particular alternative hypothesis if the power of the test tends to unity as the sample size tends to infinity; and, similarly, it has been consistent with respect to a class of alternatives if it has been consistent with respect to each member of the class.

Constant: A mathematical term for a figure that remains constant or is assumed by some hypothesis to remain constant. There are the universal mathematical constants such as pi and epsilon, conversion factors such as 'knots to feet per second, and the so-called physical constants such as the velocity of light. Special constants of interest to management statisticians include marginal costs the elasticities of demand with respect to price and to other factors and other regression coefficients.

Constraint: A constraint in a set of data has been a limitation which has been imposed by external conditions, e.g., that a number of variate values shall have zero mean, or that the sum of frequencies in a set of classes shall be a prescribed constant.

There has been another sense in which statistical data may be said to be constrained. It is the case of subnormal dispersion which was discussed by Lexis.

Consumer price index: A price index which is designed to measure changes in the cost of some specified standard of living.

Contagious distribution: Refers to a class of probability distribution of a compound kind, generally derived from probability distributions dependent on parameters by regarding those parameters as themselves having probability distributions. The name is derived from the use of such compound distributions in the study of contagious events like accidents, occurrences of disease or 'persistence' in weather.

Contaminated distribution: Refers to those distributions containing observations which have been of doubtful accurrancy or validity.

Contingency coefficient: A coefficient which attempts to measure the degree of association or dependence between two variates as expressed in a contingency table.

Contingency table: A table having R rows and C columns. Each row corresponds to a level of one variable; each column, to a level of another variable. Entries in the body of the tables, are the frequencies with which each variable combination occurred.

Continuity: A parameter or a variate is considered to be continuous when it may take values in a continuous range. A frequency or probability distribution is sometimes considered to be continuous when it relates to a continuous variate, and sometimes when the function itself is continuous. Although in statistical practice the two have been after equivalent it has been better to keep the usage clear by referring to a distribution of a continuous variateor a continuous distribution of a variate as the case may can be.

It is possible to represent a continuous, distribution by a discontinuous one by dividing its frequency into categories based s on interval measurements. Thus, for example, an age distribution, which is theoretically continuous, is frequency represented in terms of the frequencies in five-year age groups.

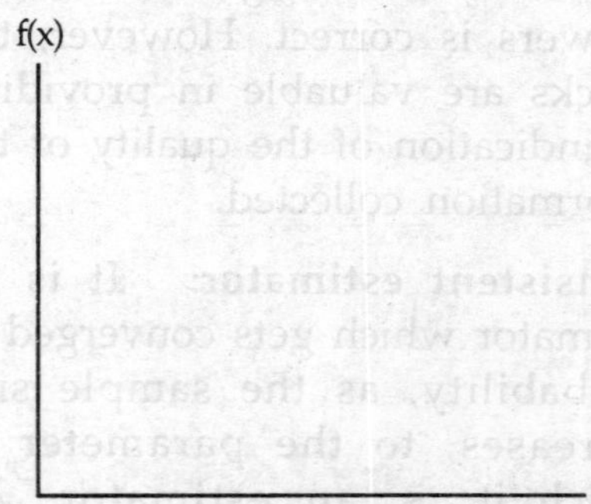

Continuous distribution

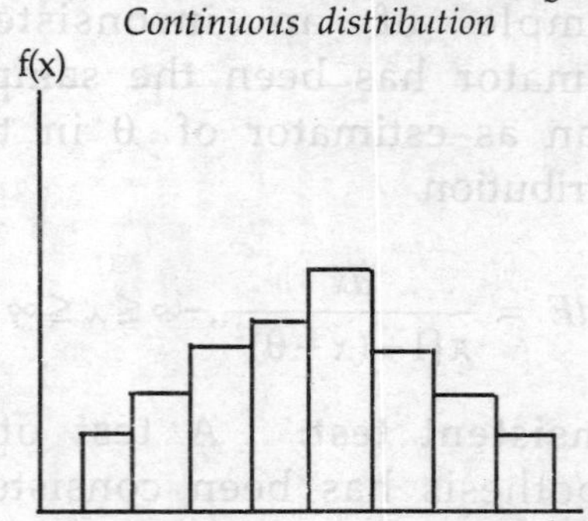

Continuous distribution

Continuous sampling: An extension of acceptance sampling and double sampling techniques. Using a given maximum number of defective or maximum rate of occurrence of effects (x) and a predetermined confidence level, it is possible to calculate an acceptance number and a rejection number for any size of sample. As the size of sample increases, so will the acceptance and rejection numbers, If at any time the number of defects is less than the acceptance number, the sample is accepted. If at any time the number of defects is greater than the rejection number, the sample is rejected. Otherwise sampling is continued.

Continuity corrections: Refers to the corrections that may need to be applied when the distribution of a discrete random variable has been approximated by that of a continuous random variable.

Continuous data: Data that may progress from one class to the next without a break and may be expossed by either whole numbers of fraction.

Continuous population. A population is sometimes said to be continuous when considered in regard to some variate which is continuous. The usage is some what not very exact, as a population may be continuous in one variate and discontinuous in another; but is permissible when the meaning is evident from the context.

Continuous probability distribution: A probability distribution in which the variable is allowed to take on any value within a given range.

Continuous probability law: probability law is said to be continuous if it corresponds to a continuous distribution function. A more stringent definition, as used in advanced probability theory, would prefix the word 'absolutely'.

Continuous process: It is sometimes used to denote a stochastic process $[x_t]$ which depends on a continuous parameter t. It is apt to cause confusion with the continuity of z and is perhaps, better avoided. The same applies of 'discontinuous' or discrete' process.

Continuous sampling plans: Refers, to a type of sampling inspection plan for use where a, production: process has been continuous in the tense that batching has been not a rational procedure. It is also applicable where immediate process control is needed and the rectification of outgoing products. The plans could be single-level or multi-level.

Continuous variable: Refers to a variable which may take any value (between reasonable limits);

this contrasts with a discrete variable, which may take only certain values. Hence the height of an adult woman has been a continuous variable,

Continuous variation: Items expressed in terms e.g., of weight, volume, length and money value, are continuous in their variation. Items expressed in terms of whole numbers are not continuous and are said to be discrete in their variation. Quantity figures of oranges sold by weight are continuous, those of oranges sold by number of units or creates are discrete.

Contrast: A contrast among the parameters in analysis of variance, or among the treatment and interaction effects in an experiment, has been a linear function of these quantities with known constant coefficients which sum to zero. It is considered to be an elementary contrast if it is having only two non-zero coefficient +1 and —1.

Control: There happen to be two principal ways in which this term finds use in statistics. If a process produces a set of data under what are essentially-the same conditions and the internal variations are found to be random, then the process is considered to be statistically under control. The separate observations have been, in fact, equivalent to random drawings from a population distributed, according to some fixed probability law.

The second usage is concerned experimentation for the testing of a new method process or factor against an accepted standard. That part of the test which involves the standard of comparison is termed as the control.

Control charts: (In quality Control) diagrams which are used to give warning of when a machine or process is not functioning well. In this context manufacturing processes Could be divided into two types. There are those whose product is either good or faulty like electric light bulbs, and others where the quality of the product is determined by measurement. If a machine for making 5 cm nails is producing them with a mean length of 4.8.cm such nails are unsatisfactory, as they do not-meet the specification; they can, however, still be used as nails, whereas faulty light bulbs are quite worthless.

In-the situation where the product has been either good or faulty', an inspection-scheme would involve recording the numbers of defective in samples of a certain size. This information could then be plotted on a control chart. This chart would behaving two lines drawn on it, the inner control (or warning) limit and the outer control (or action) limit; see Fig. The running limit is usually set so that if the process is running correctly it gets exceeded by 5% of the samples, the action limit by 0.1%.

In cases where a measurement gets involved, two things require to be watched, the mean of the variable and its dispersion. A change in either could give rise to an unduly high proportion of the products being outside acceptable tolerances limits.

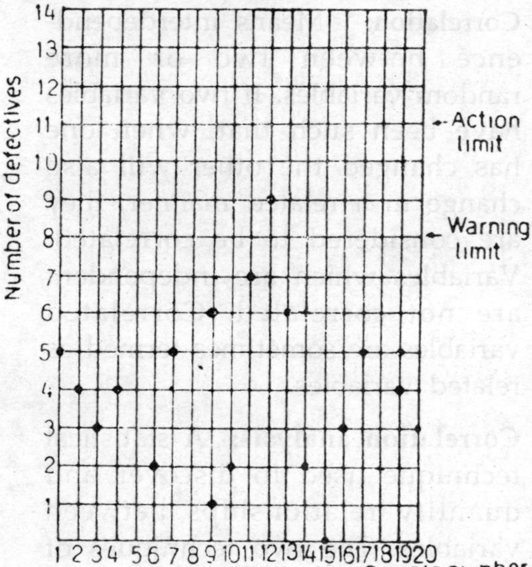

Control charts. The chart for numbers of defective in samples of size 100 (process average, 40%).

Two control charts are thus required, one for the mean, the other for the standard deviation of the samples' The control limits have been typically set so that 5% of items from a correctly functioning machine lie outside the warning limits, 0.1% outside the action limits. If the standard deviation has not been known it has been often estimated using the tables on page 257, for converting range into standard deviation.

Control charts may also at times be used for measurements from individual items, rather than for the means and dispersions of the samples.

Control experiment: An experiment which was designed to check or correct the results of another experiment by removing one of the variables operating in that other experiment.

Control group: In experimental testing a group of persons or objects used as a standard of comparison or accepted norm with which to evaluate others among which a new process or method, or set of processes and methods, has been implemented.

Controlled process: A production process subjected to valid quality control Procedures so that the size, input content or output requirements of a given product are statistically stable.

Control limits: Limits used in quality control for prescribing tolerances or allowable deviations from a required standard size in measurement or quantity in input and output. A given process may indicate that it is possible to manufacture a match with a length of, 4 cm and standard deviation a of 1 mm.

Control of substrata: A term which is used in sampling inquires to denote the employment of prior knowledge of the population cell values in an *n*-way table formed according to the *n* factors which have been used in a scheme of multiple stratification.

Controlled process: An industrial process is considered to be controlled when the mean and variability of the product remain stable. The variation has been then due to random effects or the combination of small factors of a non-cumulative kind.

Corrected moment: Refers to a moment of a set of observations which is adjusted for some effect such as the bias arising from its being calculated from a grouped frequency distributor rather than from the original data.

Correction for continuity: If a statistic is essentially discontinuous, but its distribution function is being represented approximately by a continuous function, the probability levels could sometimes be more accurately determined by entering the tables of the continuous function, not with the actual values of the statistic, but with slightly corrected values. These values are then considered to be corrected for continuity.

Correction for grouping: When data have being grouped into frequency distributions the approximation which becomes necessary by reason of having to regard frequencies as being concentrated at the mid-points of class intervals may be able to give a bias to the calculations of the moments of the distribution. Under certain conditions it becomes possible to correct for this effect, the best-known Corrections being due to Sheppard.

Correlation: Means interdependence between two or more random variables. If two variables have been such that, when one has changed the other will also change in a related manner, they are considered to be correlated. Variables which are independent are not correlated. Correlated variables are sometimes termed as related variables.

Correlation analysis: A statistical technique used to discover and quantity relationships between variables. The two techniques of correlation and regression analysis are closely related. Two variables are Correlated if a change in one variable is associated with a change in another. This need not imply a causal relationship since the movement of both variables may be influenced or controlled by other variables which may or may not have been included in the experiment or the data'. Correlation can be positive (when a positive change in one variable is accompanied by a positive change in the other), or negative (when a positive change in one variable is accompanied by a negative change in the other). The degree of correlation between two variables is measured by a statistical parameter called the covariance.

Correlation coefficient: Refers to a measure of association between two random variables. If two random variables have been such that when one gets changed the other will do so in a related manner, they are regarded to be correlated. Variables which are independent are not correlated. The correlation coefficient has been a number between—1 and 1. It could be calculated from a number of pairs of observations which are normally referred to as points *(x, y).* A coefficient of I implies perfect positive correlation, —1 perfect negative correlation, and 0 no correlation.

Correlation index: An absolescent term which refers to a quantity which purported to measure correlation where regression relationships between the variates concerned had been not linear.

Correlation surface: Refers to an almost obsolete synonym for a bivariate frequency surface.

Correlation table: Refers to the frequency-table of a bivariate distribution. The difference between a correlation table and a contingency table has been that the former generally represents a grouped frequency, distribution with intervals defined in terms of the variate values and therefore having a natural order and clearly defined width.

Cost function: In sampling theory, it refers to a function giving the cost of obtaining the sample as a function of the relevant factors affecting cost. It could be able to relate to only a part of the entire cost, e.g., by providing for the cost of collecting the sample but not for the cost of tabulation.

Cost is a measure of input. Whether input is a function of output, or output a function of input, is the kind of question that runs all through management statistics. When the method of least squares is applied, it becomes a problem as it does with mathematical model building.

Cost-of-living index: An index number of retail prices made by the Government of and published monthly. Cost- of-living index is the old name; it is now called the retail-price index or the index of retail prices. As the index is used extensively for a wide-range of purposes the Department goes to a great deal of trouble in compiling it. The formula used is

Laspeyres's index. Its pattern of-weights based on family budget inquiries, which are conducted from time to time. It has been said that this makes the index more of the Paasche type than of the Laspeyres type, but this is not so. A Paasche index is currently weighted; the retail-price index is never currently weighted: the pattern of base-period weights is merely brought up to date from time to time, a process that makes it what might be called a hybrid chain-base index. The present index includes rent, rates, other housing costs, public transport, travel, other retail services, fuel and light, meals out and holidays, as well as food, clothing, durably household goods and other goods.

Applications. As a deflator to convert money values to real values and in wages negotiations.

Countable set: It is a non-empty set, each of whose members could be labelled with a distinct positive integer. If the elements of such a set could be labelled using the integers 1, 2, 3, N, the set i s considered to be a *finite* set, with N elements. If no such N exists, the set is considered to be *countably* infinite.

The set of prime factors of the number 462 has been finite, with, 4 elements (2, 3, 7 and 11),

The set of all prime numbers has been countably infinite.

The set of all points on a plane has been infinite, but not countably infinite.

The set of all points on a plane has been infinite, but uncountable.

Counting distribution: It is a probability distribution which is formed by the number of events taking place in a fixed period of time. If the counting period starts just after an event the distribution may be known as synchronous distribution.

Covariance analysis: It is an extension of the analysis of variance to cover the case where members falling into the classes bear the values of more than one variate., interest centres on one of these (chosen as dependent variate) and the question has been whether its variation between -classes has been due to class effects or to its dependence on the other variates which themselves vary among classes. This is discussed by 'considering the -regression of the dependent variate on the other variates and the variation of the regressions (or equivalently of covariances) among classes. The technique has been found to be similar to that of variance analysis but considerable more complicated.

Covariation: Refers to the joint variation of two or more variates

Coverage: A term which has

been used in sampling in two senses (1) It denotes the scope of the material collected from the sample members (as distinct from the extent of the survey, which refers to the number of units included); (2) This term is also used to mean the extent or area covered by the sampling as in expressions such as '50 per cent coverage', which implies that one-half of the population under discussion have been examined.

Cox's theorem: A theorem given by to D.R. Cox (1952), This theorem states specific conditions under which a sequential test of a mean of a Normal distribution against a composite hypothesis could be constructed. The test has been based upon the application of the sequential probability test to transformed values of the original observations.

Craig effect: When certain short-cut methods of periodgoram analysis are used on series consisting of a random element superposed upon a scheme of harmonic components there results a bias in the estimate of the periods. The magnitude has been found to depend upon the variance of the random component.

Critical region: For the purpose of testing a statistical hypothesis at 1 per cent, 5 per cent or any other level significance, the total sample a pace is divided into regions which are mutually exclusive. Sample statistics may either fall into the region of acceptance which is usually the larger region, or into that of rejection, which is usually the smaller one. The region may consist of two areas. For example, in a two-tail test the region is, a combination of two areas, one at each extreme of the curve. The critical region is that of rejection (i.e., of non-acceptance). In the student's t-test, for example, it is represented by the areas under the curve at one or both extremes beyond a particular critical value, but this is only one well-known example of a critical region. Generally the term may be applied to the rejection region for any tests of hypothesis.

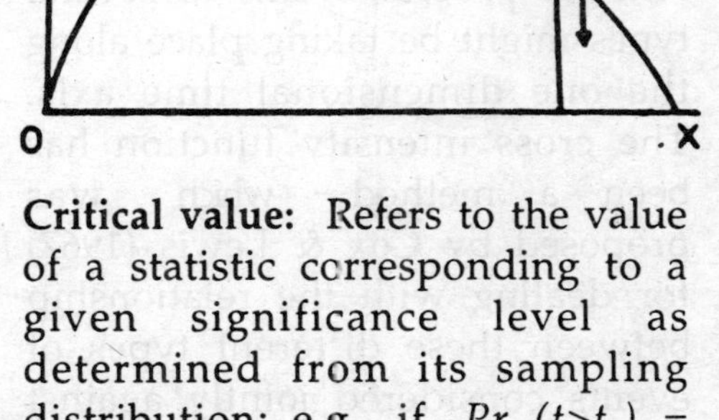

Critical value: Refers to the value of a statistic corresponding to a given significance level as determined from its sampling distribution; e.g., if $Pr\ (t>t_0) = 0.05$, t_0 has been the critical value of t at the 5 per cent level. The critical value for any test depends on the significance level at which the test is being applied, and on whether it is a one-or two-tail test.

Cross-classification: One of the simples methods of multiple regression analysis applicable to problems in up to five variables. It is, in practice, an adaptation of the simple regression method of group averages to problems in three or more variables. Although a method of multiple-regression 'analysis in its own right, it appears to be used mainly in empirical model building for determining the type of mathematical model that best fits the statistics. Where the type of mathematical model is known to be linear, a two-tier system of cross- classification can be used for a regression analysis, but where the object of applying the method is to determine the appropriate type of model, a three-tier system is necessary.

Cross intensity function: In a renewal process events of several types might be taking place along the one dimensional time axis. The cross intensity function has been a method which was proposed by Cox & Lewis (1967) for dealing with the relationship between these different types of events considered jointly against time.

Cross-section series: A term applied to any statistical series that relates to different things or different places at the same times as distinct from a time series which relates to the same thing or the same place at different times.

Sometimes hybrid series are used for analytical purposes. If, for instance, the transport firm had only five vehicles in use, then the three statistical series might relate to each of the vehicle's operations in each of three weeks, in which case the statistics could be described as hybrid.

Cross-section analysis: It is a stastical analysis of variations in data which are obtained from a set of sources at the same time. Compare time series analysis.

Cross-tabulation: A classification in a two-way table of the elements of a population according to their scores or values on two variables is called a cross-tabulation. Each cell contains the number of elements which have common scores or values on each of two-variables.

Cross-validation criterion: This criterion involves the controlled or uncontrolled division of a data sample into two subsamples. The choice of a statistical predictor, including any necessary estimation, has been made on one subsample and then the assessment of its performance has been made by measuring its predictions against the other subsample.

Crossed weight index number:

An index number is regarded to have crossed weights if it is resulted from two subsidiary index numbers, with different weights, after the application of some process of averaging.

The most commonly quoted crossed weight formula has been that of Fisher's 'ideal' index which has been the result of geometrically crossing (averaging) the index number formulae attributed to Paasche and Laspeyres. It could be written as a price index:

$$I_{on} = \sqrt{\frac{\sum p_n q_o}{\sum p_o q_o} \quad \frac{\sum p_n q_o}{\sum p_n q_n}}$$

where *o* and *n* denote subscripts relating to the base year and current year respectively. The Marshall-Edgeworth-Bowley index is also having crossed weights in this rense.

Cubodial lattice design: This experimental design, has been a development of the cubic lattice design in much the same way as the rectangular lattice has been a development of the square lattice. The cubic lattice design has been found to be suitable for numbers of treatments which are prefect cubes (k^3) whereas the cuboidal lattice design has been appropriate to a number of treatments of the form k^2 (k+ 1).

Cumulant: The cumulants have been constants of a frequency distribution defined in terms of the moments by the identity in *t*

$$\exp\left(\sum_{r=0}^{\infty} \frac{k_r tr}{r!}\right) = \sum_{r=0}^{\infty} \frac{u_r' t^r}{r!}$$

They have been thus given by the coefficients in the expansion of a power series formed from the logarithm of the characteristic function of a variable, it such an expansion exists. The earlier name for the quantities had been semi-variant or half invariant, a term introduced by Thiele. The word cumulant has been due to Cornish and Fisher (1937).

Cumulative error: Refers to an error which, in the course of the cumulation of a set of observations, does not tend to zero. The relative magnitude of the error does not then decrease as the number of observations increases.

Cumulative frequency (probability) function: A synonym for the distribution function.

Cumulative frequency distribution: A tabular display of data showing how many observations lie above or below certain values.

Cumulative frequency: The sum of all the frequency values of a given variate (x) below and inclusive of a given measure of for ordinal or interval level variates.

Cumulative frequency polygon: A frequency distribution polygon, which joins together plotted cumulative frequency values, of a given variate. If the table illustrating cumulative frequency is used, a polygon can be drawn:

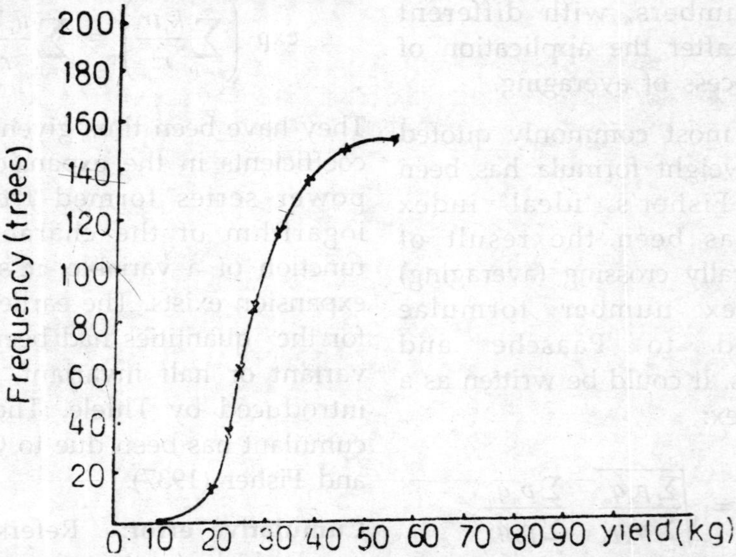

A cumulative frequency curve is a smooth continuous depiction of variate frequencies, but otherwise resembles the cumulative frequency polygon shown above.

Cumulative frequency may be expressed in the form of a precentage. For example, in the cumulative frequency polygon ditched above, the figures in brackets on the right-hand side of the vertical axis are cumulative percentage frequencies, which express cumulative frequencies as percentages of the total number of observations.

The advantage of cumulative percentage frequency curves and gives is that two or more curves depicting distributions of unequal numbers of observations may be compared using the same diagram.

Cumulative normal distribution: Refers to the cumulative frequency function (distribution function) of the Normal distribution.

Cumulative process: It refers to a development of the regenerative process, and a generalisation of the renewal process, introduced by Smith (1955). It deals with accumulation of regeneration points, or some attribute taking place at such as point, with the passage of time. An alternative formulation by D.G. Kendall (1948) presented this process as a variant of the standard birth and death process.

Cumulative sum chart: These control charts have been intended to replace the standard form, the aim has been "to make it unlikely that lack of control' will be indicated when a process has been 'in control' or that a market change in population mean will remain undetected. In cumulative sum control charts cumulative totals have been plotted against the number of observations. If $\bar{x}_j$ represent the mean of the *jth* sample, and a the known standard deviation of $\bar{x}_j$ then it would be convenient to consider the points on the control chart as having coordinates (m, X_m) where

$$X_m = \sigma^{-1} \bar{x}_1 [\bar{x}_1 - \mu) + (\bar{x}_2 - \mu) + \ldots + (\bar{x}_{ml} - \mu)]$$

and μ represents the target mean. In practice this can be achieved by calculating the cumulative sum

$$\sum_{j=1}^{m} \bar{x}_j - \mu$$

and using an appropriate scale.

Curtate: A word which is used in vital statistics to denote the integral number of years, as distinct from the neatest number of years, for which a given state has existed. For example if an assurance matures in 3 years 9 months, the curtate duration would be 3 years. An analysis which uses curate periods would, in most cases, be subject to downward bias.

Curve fitting: An expression which is being used in two rather different senses in statistics: (a) In one sense, it is used to denote the fitting of a mathematically specified frequency curve to a frequency distribution (cf. Pearson curves); (b) In the second sense, it is used to denote the fitting of a mathematical curve to any statistical data capable of being plotted against a time or space variable, *e.g.*, regression data or time series.

Curvilinear correlation: An expression which is used to denote correlation in bivariate data when the regressions are not linear. Strictly' speaking it has been a misnomer, a correlation being a number and not a function admitting curvature, and has been perhaps better avoided.

Curvilinearity: A relationship between two or more variates which is depicted graphically by any locus other than a straight line. For example in time series analysis the accelerating rate of increase in prices is called a curvilinear trend and represented by a steepening curve of prices plotted against time. Curvilinearity could represent a logarithmic, quadratic or higher order relationship, or a more complex function.

Curvilinear regression: It is a type of regression which is not linear. A form often considered has been that for which the dependent variate has been expressed as polynomial in the independent variables.

Curvilinear relationship: An association between two variables that is described by a curved line.

Curvilinear trend: Refers to a trend which is not linear. It could be expressed as a polynomial, a more complicated mathematical expression such as a logistic curve, or by some smoothing process like a moving average.

Cut-off: Refers to the artificial truncation of a sampling process at a point when it becomes apparent that enough data are being collected for the purpose in view.

Cyclical movements/fluctuations: In time series analysis cyclical movements are sequential, repetitive, but not necessarily regular, movements on either side of a given trend line. The object of time series analysis in to discover the extent to which a time sequence of variable values may be predicted by: (1) a trend line; (2) cyclical movements; (3) episodic movements caused by unusual events; and (4) residual random fluctuations. Thus cyclical movements are oscillatory movements superimposed on the trend line in wave like fashion. In business statistics, for example, they may include both (a) season variations and (b) variations resulting from 'trade cycle' conditions, although the definition is usually reserved for (6) in some statistical texts and seasonal fluctuations are isolated separately.

In other branches of statistics, migratory cycles, political cycles (e.g. between Presidential elections in the United States) fall into the same category.

Cyclic component: It is a non-seasonal component in a time series which varies in a recognizable cycle. An example is the frequency of sunsports, which has an 11.2-year cyclic component.

Cyclic design: Refers to a class of balanced and partially balanced experimental design in which the blocks have been constructed by cyclic permutations of the treatments.

D

Data: The plural of the Latin word 'datum'. The word can mean any information which is 'given' or provided for the solution of a problem. Because more than one unit of information is necessary for solving every problem, the word is plural. Data must be quantified and structured before being usable for the solution of a statistical problem. They are measured in statistical units, which must be: (1) relevant to the enquiry or problem; (2) homogeneous (*i.e.*, of the same type) and (3) stable.

In accounting statistics a measure of homogeneity if often achieved for comparing dissimilar units, for example, the standard hour, a measure for work which a typical individual performs in an hour.

Data are often divided into two categories, primary and secondary indicating nearness to source or otherwise. For example, data collected by means, of experiment,interview, observation, questionnaires or survey are primary, while items of data obtained by the researcher from a previous source or publication are secondary.

Data array: The arrangement of raw data by observations in either ascending or descending order.

Data base: A store of a large amount of information which is specially in a form that could be handled by a computer. Information in a data base has been stored as a number of records or files, each of which usually has entries under a number of headings of fields.

Data capture: Any process which is used for converting information into a form that could be handled by a computer.

Data point: A single observation from a data set.

Datum zero: A base level from which values of a, variable have been taken. For example, heights of tides have been given with respect to a datum zero, called Low Water Ordinary Springs, (LW), and defined physically by a mark on Newlyn harbour wall; dates are measured with respect to a datum zero, the start of year 1 AD; longitude is measured with respect to the Greenwich meridian as datum zero in the assumed mean method of working out averages the assumed mean has been a datum zero for the numbers involved.

David's empty cell test: It is a distribution-free test. According to this test a continuous popul-ation possessing a particular completely specified probability density function against the general alternative hypothesis that it has not. The idea behind the test has been that if the area under the hypothesised density function has been divided into a number of vertical strips of equal area, the bases of which have been regarded as cells, then a simple random sample from the actual population would tend to be more evenly dispersed among the cells if the hypothesis were true than if it were false. The number of empty cells should thus give rise to a reasonable index of the truth or falsity of the hypothesis.

D^2 statistics: A statistic which was introduced by Mahalanobis and is regarded as a measure of the 'distance' between two populations with differing means but identical dispersion matrices. The distance between the populations

$$\Delta^2 = \Sigma a_{jk} (\mu_{1j} - \mu_{2j}) (\mu_{1k} - \mu_{2k})$$

is termed as Mahalanobis' generalised distance. The sample value of Δ^2 would be as follows

$$\Delta^2 = \Sigma\, a_{jk}(\bar{x}_{1j} - x_{2j})\ (\bar{x}_{1k} - x_{2k})$$

and (a_{ik}) would be the inverse of the pooled dispersion matrix. Developments of this concept, principally by the Indian school of statisticians, have been sometimes collectively termed as the'distance method'.

Death process: Refers to a stochastic process in continuous time defined on the state space of non-negative integers, representing the decay of a population in which only deaths may occur. The expression is generally confined to processes for which instantaneous transitions to the next lower integers only are possible, the probability of a transition from n to $n - 1$ time dt being asymptotically $\mu_n dt$; μ_n might be time dependent.

Death rate: It may be defined as the number of deaths in a given period divided by the population exposed to risk of death in that period. For human populations the period is generally one year and if the population is changing in size over the year the divisor is considered as the population at the midyear. The death rate as so defined is termed as 'crude'. If some refinement has been introduced by relating mortality to the age and sex constitution (or other factors) for comparative purposes the rate will be considered to be standardised.

Decapitated negative binomial distribution; It is a negative binomial distribution from which the zero class is missing. The term 'decapitated' was used by Yule to characterise a truncated distribution from which the zero class only is removed.

Decile: One of a series of partition values, known as quantities, which divide the frequency of a group, set, sample or population into equal proportions. As their name implies, deciles divide the total frequency of a set of variate values into ten equal partition.

Decimal places: Refers to the figures to the right of the decimal point that are needed to press the magnitude of a number to a specified degree of accuracy.

Decision function; It is a rule of conduct which, at any stage of a sampling investigation, informs the statistician whether to take further observations or whether enough information has been collected, and in the latter case, what decision to make upon it. At each stage beyond the first the decision function has been a function of the preceding observations.

Until the development of sequential methods decision functions had been mostly of the simple type, based on a fixed sample size, which enjoined the acceptance or rejection of a hypothesis or set limits to a parameter under estimate. The above definition gives a sequential situtation wherein the investigator may not able to reach a, decision about the hypothesis but proceeds to take further observations.

Decision space: In sequential analysis and the theory of decision functions, the decision space has been the set of all possible decisions.

Decision tree: A statistical device for estimating the optimum of order for Perishable goods, or of a plant for making any kind of standard goods. As originally depicted, the decision tree consists of a number of main branches, each representing a size of order or plant and each with one or more secondary branches representing judgements of the financial consequences.

The main branches grow out of a box labelled D (for decision) situated on the left of the picture and extending to the right, the secondary branches extending further to the right, and the whole tree lying prone as though it had just been felled. In its most logical and practicable form, the decision tree exploits the frequency distribution. The decision tree is not to be confused with the economic ordering quantity which is concerned with stores control through ordering frequency.

Decomposition: Refers to the act of splitting a time series into its constituent parts by using statistical methods. A typical time series in often considered as composed of four parts:

(a) A long-term movement or trend;

(b) Oscillations of more or less regular period and amplitude about this trend;

(c) A seasonal component

(d) A random, or irregular, component.

Any particular series should not exhibit all three of these but those which are present are assumed to act in an additive fashion, *i.e.*, are superimposed; and the process of determining them separately is one of decomposition.

A more modem approach needs to decompose the series into deterministic and non-deterministic elements. This is termed as Wold's decomposition or predictive decomposition.

Decreasing hazard rate: A decreasing hazard rate may be defined as one which does not increase; it can, therefore, be constant.

Deep stratification: A term used to denote stratification by a substantial number of factors with respect to the marginal distribution of the factors only.

Defective probability distribution: A probability distribution usually assigns a probability measure over the appropriate interval in such a way that the sum of all the assigned probabilities would be unity. However, probability measures have been sometimes encountered which assign a total mass, say *p*, which has been less than unity. Such a measure is termed as a defective probability measure with defect 1–*p*: the probability distribution has been a defective distribution.

Defective sample: A sample which is incomplete or results from an incomplete survey or examination. Two examples of defective samples are;

1. a sample taken from a telephone directory in an enquiry related to the study of wealthy people. Although it is valid to assume that most people whose incomes are above a given level have telephones, some of the most wealthy people are x-directory.

2. a sample which is incomplete, because some records or completed questionnaires have been lost.

Defective sample: It is a sample which has resulted from an inquiry which has been incompletely carried out, e.g., because certain assigned individuals are not examined, because records are lost, or because certain members are dead.

Defective unit: In quality control, this term refers to a unit which does not reach some prescribed standard and therefore gets, rejected, in contrast to an effective unit.

De Finettis theorem: This theorem is due to de Finetti. According to this theorem, a simple sample that generates a permutable binary sequence can always be considered as having been obtained by binomial sampling in which the type I probability P has a Type II initial distribution which is unique.

Defining contrast (or relation): In the analysis of half-replicate factorial designs the comparison between those treatment Combinations which are used with those that are not used is termed as the defining contrast. It defines the halves into which the whole replicate gets divided.

Deflation of values: An economic time series measured at current' prices can be converted to a constant price series, giving a better indication of changes in volume by dividing it by an index of the prices of its components. This technique is referred to as deflating the series. Some difficulties involved are the choice of base period in terms of whose prices the deflated series is expressed and the fact that the relative importance of the components of the series will change as their relative prices change. This means that a suitable index of prices often cannot be Constructed and a proxy measure must be used such as a retail or wholesale price index.

Degenerate distribution: Refers to an alternative designation for deterministic distribution.

Degrees of freedom: This term has been use in statistics in slightly different senses. It was introduced by Fisher on the analogy of the idea of degrees of freedom of a dynamical system, that is to say the number of independent coordinate values which have been necessary to determine it.

In this sense the degrees of freedom of a set of observations (which *ex-hypothesis* are subject to sampling variation) has been the number of values which could be assigned arbitrarily within the specification of the system. For example, in a sample of constant size n grouped into k intervals there have been $k - 1$ degrees of freedom because, if $k - 1$ frequencies are specified, the other has been determined by the total size n; and in a contingency table of p rows and q columns with fixed marginal totals there are $(p - 1)(q - 1)$ degrees of freedom.

Degrees of randomness: Strictly speaking there have been no degrees of randomness, except in the theory developed by M.G. Kendall under which randomness may be defined as relative to an assigned test domain. The expression is, however (rather regrettably) introduced into the theory of stochastic processes.

A random function $\phi(t)$ possesses n degrees of randomness if n is the smallest integer such that if n values are known at distinct points of time then it is almost certain that a functional relationship exists between $\phi(t)$ $[T>t_n]$ and these available values.

Demand schedule: A statistical schedule designed to show the effect of changes in price on the quantity of a brand of product sold in unit time, and on the sales proceeds.

Demodulation: A technique which was used in spectrum analysis of time series for the detecting and describing the more unusual kinds of non-stationary behaviour. The method involves multiplying a series by a suitable function and applying a Filter to leave only the frequency in a narrow bank.

Demography: The study of population size, composition, and distribution, and the patterns of change therein. The narrowest conception of demography views it as the study of vital statistics.

De Morgan's laws: The laws which state that, for sets or events $A_1, A_2, \ldots A_n$,

$$\text{a)}\quad \left(\bigcup_{i=1}^{n} A_i\right)' = \bigcap_{i=1}^{n} Ai'$$

$$\text{b)}\quad \left(\bigcup_{i=1}^{n} A\right) = \bigcap_{i=1}^{n} Ai$$

These may be stated in words as:

(a) The complement of the union of events is the same as the intersection of their complement.

(b) The complement of the intersection of events is the same as the union of their complements.

These laws could be illustrated by this simple example from sets.

$$\xi = \{a, b, c, d, e, f\},$$

$$A_1 = \{a, b, c\} \text{ and } A_2 = \{b, c, d\}.$$

In this case, *de Morgans a first law* gives

$A_1 \cup A_2 = \{a, b, c, d\}$, and so $(A_1 \cup A_2)' = \{e, f\}$

$A_1' = \{d, e, f\}$, $A_2' = \{a, e\ f\}$, and so $A_1 \cup A_2 = \{e, f\}$

so $(A_1 \cup A_2)' = A_1' \cup A_2'$

de Morgan's second law gives

$A_1 \cap A_2 = \{b, c\}$, and so $(A_1 \cap A_2) = \{a, d, e, f\}$

$A_1' \cap A_2' = \{a, d, e, f\}$

so $(A_1 \cap A_2) = A_1 \cap A_2'$.

Departure process: It is the stochastic process which is formed by the intervals between the departure times from the service counter or station of a queuing situation. This departure process has been of importance in the study of tandem queues.

Dependent samples: Samples drawn from two Populations in such a way that the elements were not chosen independently of one another, in order to allow a more precise analysis or to control for some extraneous factors.

Dependent variable: A variable which depends on another (the independent, or explanatory variable). If we draw a graph, the dependent variable is generally drawn on a vertical (y) axis the independent variable on the horizontal y axis.

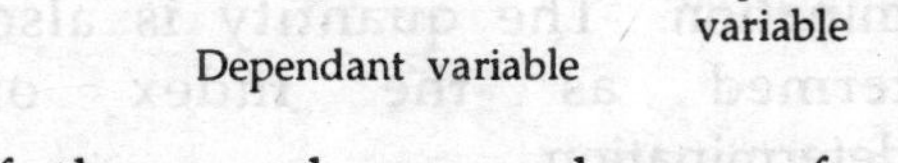

Dependant variable

If the graph were drawn of a girl's height, measured every birthday, against her age, her height would be the dependent variable, her age the independent. Her height depends on her age but her age does not depend on her height, just on the length of time she has been alive.

Derived statistics: This type of statistics is obtained by an arithmetical operation from the primary observations in this sense, almost every statistic could be derived. The term is especially used to denote descriptive statistical quantities which are obtained from data and are primary in the sense of being mere summaries of observations, *e.g.*, population figures have been primary and so have been geographical areas but population per-square-mile is a derived quantity.

Descriptive indices (Indici Descrittivi): An index (or coefficient) is called descriptive if it shows a relationship concerning each value of the variable, as distinct from the global or mean index or coefficient (*indice global or medio*) which relates only to the aggregate or the mean of the values. Means, standard deviations, mean differences, etc., have been global indices; Pareto's a and Gini's δ for incomes purport to be descriptive indices. An index, which has been descriptive for one characteristic, may be only global for another: the a and δ indices have been descriptive for incomes, but only global for fortunes.

Descriptive statistics: As statistics, has two distinct meanings (1) the plural of statistic and (2) methods of collecting, presenting and analysis data, so descriptive statistics can mean: (a) statistical data of a descriptive type, or (b) simply calculated measures of data, which are of a descriptive rather than an analytical type. The calculated statistics generally relate to the dala on which they are based, rather than to a population from which the data may have been drawn.

Most writers on statistics confine the term to: (i) methods of presenting statistics, using tables, graphs, charts, pictograms etc; (ii) some simple methods of calculating measures of central tendency (*e.g.*, mean, median and mode) and measures of dispersion *e.g*,, standard deviation).

Deseasonalization: A statistical process used to remove the effects of seasonality from a time series.

De-seasonalized series: A time series the observations of which have been adjusted by removing the variation due to seasonal factors (the seasonal variation).

Design of statistical: A statement of the purpose of, and proposed approach to an experiment or investigations involving statistical analysis. The approach covers the statistical methods to be used and a realistic form of mathematical model, which takes account of the relevant enable basic and derived statistics.

Destructive test: Under certain conditions it has been possible that the carrying out of an inspection test on a manufactured product will cause destruction of the particular test specimen. This is termed as a destructive test. For example, a test of the fusing current in an electric has been of such a kind and many of the inspection tests for explosive products must necessarily be destructive. Under such conditions there has been every incentive to design sampling schemes which minimise the number of items to get tested.

Determination, coefficient of: Refers to the square of the product-moment correlation between two varieties, r^2, so-called because it represents the proportion of the variance of one variate, y, given by the other, x, when y is expressed as a linear regression on x.

More generally, if a dependent variate is having multiple correlation R with a set of independent varieties R^2 is termed as the coefficient of, determination. The quantity is also termed as the index of determination.

Deterministic distribution: Refers to a degenerate-distribution which consists of a mass concentrated at a single point $x = k$ so that its probability function has been

$$\Pr(x) = \begin{cases} 1 & x = k \\ 0 & x \neq k \end{cases}$$

Deterministic model: A deterministic as model, as opposed to a stochastic model, may be defined one which contains no random elements and for which, therefore, the feature course of the system has been determined by its position, velocities, etc., at some fixed point of time.

Deterministic process: Refers to a stochastic process having a zero error of prediction; one in which the past completely ascertains the future of the system.

Deviance: A term which was proposed by M.G. Kendall to and used to denote the sum of squares of observations about their mean.

Deviate: Refers to the value of a variate which is measured from some standard point of location, usually the mean. It is generally understood that the value could be expressed in standard measure, *i.e.,* as a proportion of the parent standard deviation.

Deviation: Refers to the difference between a value of a variable and the mean of its distribution.

Example: The numbers 3, 6, 7, 10 and 4 are having mean 6, and so their deviations are:

$$3 - 6 = -3$$

$$6 - 6 = 0$$

$$7 - 6 = 1$$

$$10 - 6 = 4$$

$$4 - 6 = 2$$

The sum of all the deviations in such a set should be zero. Deviation has been used in the calculation of variance, standard deviation, covariance, etc., but has been not often used in its own right.

Diagonal regression: If two variables are subject to errors of observation and get connected by a linear relation, a set of observed values will not, in general, lie on a straight line. The linear relation is estimated by regression techniques; and this method is having the embarrassing property of yielding two different regressions. Frisch gave compromises between these two in the form of orthogonal regression and diagonal regression. The latter has been a line which is lying between the regression lines and passing through section, and having the equation their point of inter section and having the eqaution

$$\frac{Y - m_y}{\sigma_y} = \frac{X - m_x}{\sigma x}$$

where $m_{y'}$ m_x denote the means add σ_y^2, σ_x^2 the variances of y and x respectively. If $\sigma_y = \sigma_{x'}$ the diagonal regression coincides with the orthogonal regression. Neither, strictly speaking, has been a regression.

Diagrammatic presentation: Presentation of statistics, by chart on a wall for general information, or in a book or article to illustrative an argument.

Dichotomous population: In statistics a population that has been divided into two mutually exclusive classes. Thus all cases in ore category have a certain characteristic or more than a certain magnitude of a characteristic and no cases in the other category have that characteristic or that magnitude, for example, a population divided into male and female, or over 21 years of age and under 21.

Difference and sum of two or More Random Variables. Refers to those variables which have been the difference and the sum of two or more random variables. If two random variables, X_1 and X_2 are added or subtracted to produce a third random variable, its mean and variance would be given by these relationships.

$$\text{Mean } (X_1+X_2) = \text{Mean } (X_1) + \text{Mean } (X_2) \text{ or}$$

$$E(X_1+X_2) = E(X_1) + E(X_2)$$

$$\text{Variance Var } (X_1+X_2) = \text{Var}(X_1) + \text{Var}(X_2) + 2\text{ cov } (X_1+X_2)$$

$$\text{Mean } (X_1\text{-}X_1) = \text{Mean } (X_1) - \text{Mean } (X_2) \text{ or } E(X_1 - X_2) = E(X_1) - E(X_2)$$

$$\text{Variance Var } (X_1 - X_2) = \text{Var } (X_2) + \text{Var } (X_2) - 2\text{cov } (X_1, X_2)$$

where E denotes expectation, cov denotes covariance and Var denotes variance. If X_1 and X_2 have been independent variables then cov $(X_1, X_2) = 0$. In that case

$$\text{Var } (X_1 + X_2) = \text{Var } (X_1) + \text{Var } (X_2)$$

$$\text{Var } (X1 - X_2) = \text{Var } (X_1) + \text{Var } (X_2)$$

Difference sign test: When observations are considered on a phenomenon which a moving through time and generates an ordered set, *i.e.*, a time series, one useful test against trend, etc., would be the difference sigh test. This consists of counting the number of positive first differences of the series, that is to say, the number of points where the Series increases. It has been used especially as a test against linear trend and as such has been superior to the Turning Point test but inferior to tests based on rank order.

Diffusion index. A term which was proposed by Burns and Moore to denote the proportion of a set of time series in a given collection of series which have been increasing at a given point of time.

Diffusion process: A type of Additive Process which describes certain kinds of diffusion The process has been such that the 'displacement' of the variate (its increment) in time dt follows a Normal distribution with variance proportional to dt.

Dimroth-Watson's distribution: It is a spherical distribution of the symmetric girdle type which was introduced independently by Dimroth and Watson. Its probability density function would be as follows :

$$f(l,m,n)=\{b(K)/2\pi \exp, \{-K(1\lambda + m\mu + m\nu)^2\}' \quad -\infty < K < \alpha$$

where (l, m, n) and (λ, μ, ν) denote the vectors of direction cosines of a three-dimensional directional random variable and its mean direction respectively, and

$$b(K) = 1/\left(2\int_0^2 e^{-kt_2} dt\right)$$

Direct probability: Refers to an expression which is assumed to be antithetical to inverse probability, although neither expression has been, very logical. It usually represents probability when used to proceed from the given probabilities of prior events to the probabilities of contingent events; *e.g.*, if it has been given that the probability of throwing each number with an ordinary six faced die has been 1/6 the probability of throwing a score of 15 in three throws is directly ascertainable. There relation has been analogous to the deductive relations of logic.

Direct sampling: This term is used when the sample units have been the actual members of the population and not, for example, some kind of record relating to such numbers, like census form, ticket or registration card. The term has been related to the directness of the observation of the units which enter into the sample, not to the process by which they have been selected.

Dirichlet distribution: Refers to the k-variate analogue of the beta distribution.

Disarray, coefficient of: Two rankings A and B may get transformed one into the other by a successive interchange of pairs of neighbouring items in one of term. There would be a minimum number of moves in which the operation could be made and a coefficient of disarray could be constructed from it. This coefficient has been in fact the same as Kendall's tau.

Discontinuous variate: Refers to a variate which may take only a discontinuous set of values.

Discrepance: In variance analysis, this term has been used by some writers to denote the error sum of squares.

Discrete data: Data that do not progress from one class to the next without a break. *i.e.,* where classes represent distinct categories or counts and may be represented by whole numbers.

Discrete lognormal distribution: It is a confusing term which is used to denote a distribution obtained as the compound of a Poisson distribution with the parameter A distributed log normally.

Discrete normal distribution: It is a discrete analogue of the continuous normal distribution of the form given below:

$$\Pr(x) = A\exp\left\{-\frac{(x-m)^2}{2\nu}\right\}$$

where $x = 0, \pm 1, \pm 2, \ldots,$ and $m \approx \mu$ with $\nu = \sigma^2$ and

$$1/A = \Sigma\exp\left\{-\frac{(x-m)^2}{2\nu}\right\}$$

with ν and $m^2/\nu \geq 10$.

Discrete pareto distribution: It is the discrete analogue of the continuous Pareto law:

$$\Pr(x) = \frac{1}{\zeta(s)}\frac{1}{x^s}\ x = 1, 2, \ldots \infty \text{ and}$$

where ζ denotes the Riemann zeta function.

Discrete power series distribution: It is a general probability distribution of the form given below

$$\Pr(x) = \underset{x}{a}\,\theta^x / f(\theta) \text{ with } x = 0, 1, 2, \text{ and}$$

By appropriate choice of $f(\theta)$ it is possible to derive several common discrete distributions; for example, the Poisson, negative binomial, Polya-Eggenberger.

Discrete probability distribution: A probability distribution in which the variable is allowed to take on only a limited number of values.

Discrete probability law: A probability law is termed as discrete if it corresponds to a discrete distribution function.

Discrete random variable: A random variable that is allowed to take on only a limited number of values.

Discrete rectangular distribution: Refers to a distribution (of finite range) in which each of a discrete set of x variables, n in number, is having the same probability $1/n$. It has been mostly used for equi-spaced variable values, e g., the numbers 0-9 if equally probable, from a discrete rectangular distribution.

Discrete series: This term is used in the literature in two senses: (1) a series which is naturally discrete, i.e., one whose values fall naturally at a number of fixed exact measurements., *e.g.,* a number of fixed integers; and (2) a series which is expressed in discrete form by using class measurements, i.e., class intervals.

Discriminatory analysis: For a set of multivariate observations on samples, known with certainly to come from two or more populations, the problem has been to set up some rule which would allocate further individuals to the correct population of origin with minimal probability of misclassification. This problem and sundry elaborations of it would give rise to discriminatory analysis.

Dishonest process: It is an alternative name for a logical process, and the opposite in a mathematical sense, of an honest process.

Disjoint, aq. (of sets) non-overlapping: In the Verm diagram (Fig.) the sets A and B are disjoint, $A \cap B = \phi$

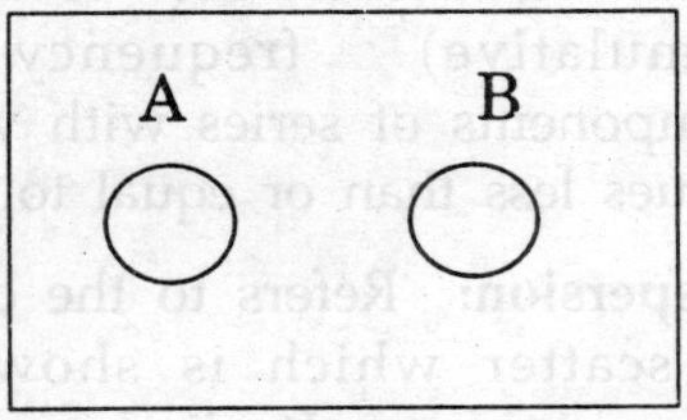

Disjoint: A Venn diagram of disjoint sets A and B

The equivalent term for events (*i.e.*, describing two events that cannot both occur) has been exclusive.

Dispersion: Refers to the spread of a distribution. In the diagram distribution B is having greater dispersion or spread than A although they both are having the same mean.

Dispersion: Diagram of two distribution with the same mean, but different dispersions. There have been several measures of dispersion. Mean absolute deviation, standard deviation and semi-interquartile range have been all measures of how far above or below the middle a typical member might be found.

Standard deviation has been the most important statistically and also, for small data sets, the most tedious to calculate by hand; it gives more weight to extreme values than do the other two.

Disproportionate allocation: Occurs in stratified sampling whenever different sampling fractions are used in selecting the sample indifferent strata. The most commonly used form of disproportionate allocation is optimal allocation.

Disproportionate stratified sampling: Stratification denotes the division of the population into subdivisions, known as strata, in each of which sampling is carried out Independently. Once the strata have been formed, the investigator' has complete freedom to choose the sampling fraction (or proportion of the elements to be selected) in each stratum. Disproportionate allocation (using different sampling fractions in different strata) will be justified on three counts: (1) the more dispersed or variable the elements are in a stratum the higher the sampling fraction; (2) the more expensive it is to obtain information in a stratum the lower the sampling fraction; (3) the more precision we require for the estimates in a stratum the higher the sampling fraction.

Disproportionate allocation does, however, involve some risks. Whereas proportionate allocation cannot in general lead to a fall in precision compared to simple random sampling, disproportional allocation—if based on faulty information about costs or dispersion—can lead to serious losses in precision. The additional costs in the analysis caused by the additional complexity of the sample design should be considered when deciding whether to use proportionate or disproportionate allocation.

Distribution: Refers to the set of frequencies *(observed distribution)* or probabilities (theoretical *distribution) which* are assigned to a set of events.

Distribution curve: The graph of a distribution function, showing frequency values plotted against an ordinate and variate values against an abscissa, The term is popularly used to denote a cumulative frequency curve.

Distribution-free: Methods. Methods of testing given hypothesis, which do not depend on assuming a particular parametric frequency distribution, but which are applicable to all frequency distributions.

It is possible to conduct tests involving ranked observations from elements of a sample, whose results would be valid quite irrespective of the parameters of the population from which the sample is drawn. Such methods of testing are distribution-free. The term non-parametric methods is sometimes used interchangeably with distribution-free methods.

Distribution function: Although the general meaning of this term may include any frequency distribution, where a given variate x, for example, may have a frequency 5 at x_1, 10 at x_1, 12 at x_3, etc., the term is often specifically used to mean the total (or cumulative) frequency of components of series with variate values less than or equal to x_1.

Dispersion: Refers to the degree of scatter which is shown by observations, It is generally measured as an average deviation about some central value *(e.g.,* mean deviation, standard deviation) or by an order statistic (e.g., quartile deviation, range) but could also be a mean of deviations of values among themselves *(e.g.,* Gini's mean difference and also standard deviation).

Dispersion stabilising transformation: A term introduced by Ruben to denote—a generalisation of the variance stabilising transformation. The generalisation covers the stabilisation of covariance matrices.

Dispersion, statistical: The degree of variation of the values of a variable from a central point or typical value, such as the mean. Measures that describe dispersion, alternately called scatter or spread, include mean deviation, quartile deviation, range, and standard deviation.

Displaced poisson distribution: A situation in which the number of events in excees of a threshold value *r*, when it is regarded that at least the *r* events do occur, is having a displaced Poisson distribution with parameters λ and *r*.

$$\Pr(x) = \frac{e^{-\lambda}\,\lambda\,(x+r)}{I(r,\lambda)(x+r)}$$

Distance: This word is finding use in many statistical contexts in its ordinary sense, e.g., the 'distance' of a value x from some origin a is *x—a*. A specialised use is found in the notion of 'distance' between two variates, x and y, which is defined as the expected value of *x—y*; or the 'distance' between two populations, which is defined as the difference of their means.

Distance distributions: Refer to distributions which arise from more or less naturally defined geometrical systems of variation. For example, if two points have been taken at random within a circle, with unit radius and centre at the origin, the joint distribution of the coordinates (X_1, Y_1) (X_2, Y_2) would be as follows:

$$\Pr x_1, y_2; x_2, y_2^{(x_1, y_1; x_2, y_2)}$$

$$= \pi^{-2}(0 \le x_j^2 + y^2 + \le 1; j = 1, 2)$$

Distance measure: A measure, of dispersion in terms of the difference between two values in the data set.

Distribution curve: It refers to the graph of cumulated frequency as ordinate against the variate value as abscissa, namely the graph of the distribution function.

Distribution-free method: It is a method, e.g., of testing a hypothesis or of setting up a' confidence interval, which does not depend on the form of the underlying distribution; for example it is possible to obtain confidence intervals for the median, based on binomial variation, which are valid for any continuous distribution. In the case of testing hypotheses, it means that the test is independent of the distribution on the null hypothesis. Distribution-free inference or distribution-free tests are sometimes termed as non-parametric but this usage has been confusing and should be avoided. It has been better to confine the word 'non-parametric' to the description of hypotheses which do not explicitly make an assertion about a parameter.

Distribution function: Refers to the function F(*x*) which gives the cumulative frequency or the cumulative probability of the random variable X. Thus, F(*x*) = Frequency (or probability) of X ≤ x for a set of values of X.

Distribution of run length: Refers to the distribution of lengths of runs of attributes in a series. A run may be defined as a sequence of items of the same kind which get terminated by one or more items of another kind.

Disturbed harmonic process: It is a form of stochastic process which was proposed by Yule to explain the continual change of amplitude and shift of phase which appears to be typical of time series from the economic and meteorological sciences. If a series has been observed at points $t = 1, 2,...$ an ordinary harmonic movement could be expressed by the equation:

$$u_{t+1} - 2u_{t+1} + u_t = 0$$

If the zero is replaced by a random term ε_{t+2} on the rightband side the motion is said to be that of a disturbed harmonic. It has been a limiting case of an auto-regressive process but has been not stationary.

Dixon's statistics: Refers to a class of statistics for the rejection of the largest observation proposed by Dixon of the form

$$r_{ij} = \frac{x_{(n)}\ x_{(n-i)}}{x_{(n)} - x_{(j+1)}} \qquad \begin{matrix} i = 1, 2 \\ j = 0, 1, 2 \end{matrix}$$

where $x(k)$ implies the kth value of a sample on n when ranked by size.

Domain of study: A domain is a part of the population in which the investigator has a special interest and for which separate estimates are desired. Stratification where possible, may be used to separate the domains of study so that the desired precision can be built into the survey design. Although the term should be used to denote subdivisions of the population for which estimates with known precision are planned, it is sometimes used to specify subdivisions for which seated but for which the re-estimates are to be calculate prior information for the design in not available.

Dominating strategy: Refers to a strategy which has been in no respect worse than all others and in some respect has been better.

Double binomial distribution: This distribution has been a weighted average of two binomials having parameters p_1 and p_2; $p_2 < p_2$ and weights w_1 and w_2; $w_1 + w_2 = 1$

Double confounding: This term means confounding of two different groups of treatment contrasts having two different sources of variation in an experimental design. Thus in the row and column layout of a quasi-latin square certain interactions get confounded with rows, and certain others with columns.

Double exponential distribution: It is the distribution having frequency function $f(x)=ae^{b/x-c|}$, $-\infty\leq x\leq\infty$, a, b and c being constants, $b<0$. The distribution may be considered to be an ordinary exponential together with its reflection about the point c. An example of the double exponential distribution has been that of the distribution of the mid-range from a rectangular population.

Double exponential regression: It is an expression which is sometimes used for a regression equation in which the dependent variable (y) has been that of the combination of two simple exponential terms in (x)

Double hypergeometric distribution: If, from a finite population having two kinds of item, a random sample has been drawn without replacement and then a second random sample without replacing the first, the distribution of the (x_1, x_2) successes would take the form of the double hypergeometric.

Double poisson distribution: It is a distribution in which the parameter A of the Poisson series is itself considered as distributed in the Poisson form. The distribution arises particularly in ecology where the numbers offspring of parents which have been themselves distributed over space in the Poisson form have been also distributed in that form in the neighbourhood of the parent.

Double-ratio estimator: Refers to an estimator which is built up from four variates, x_1, x_2, y_1, y_2 by using the ratio of the ratios y_1/x_2 and y_2/x_2.

Double reversal design: Refers to an extension of the switchback or reversal design to the case of two treatments over four periods: A_1 B_2 A_3 B_4 for one-half of the experimental units and B_1, A_2 B_2 A_4 for the other half; the suffixes have been the time periods. This design permits trends in the responses which are independent of the treatment effects.

Double sampling: It refers to a standard form of sample design for industrial inspection purposes. According to the characteristics of a particular plan, two samples have been drawn, n_1 and n_2, and the first sample inspected. The batch can then be accepted or rejected upon the results of this inspection or the second sample be inspected and the decision made upon the combined result.

Doubled-tailed test: It is a test for which the region of rejection includes areas at both extremes of the, sampling distribution of the test function. It is general but not essential to allot one-half of the probability of rejection to each extreme, yielding a symmetrical test.

Doubly stochastic poisson process: It is a point stochastic process where the rate of occurrence in a Poisson process has been itself a stochastic process.

Doubly truncated normal distribution: It is a distribution of normal (Gaussian) form but with the variable not extending to infinity, being truncated below and above at specified points.

Downward bias: It refers to a bias which tends to reduce a magnitude below its true value. In index number theory the expression is frequently found but has a rather obscure meaning owing to the somewhat arbitrary nature of the definition of bias in index numbers.

Dual process: For every queueing process it has been possible to Create an associated process by interchanging the distributions of inter-arrival and service times respectively: this associated process is known as the dual process by Takacg.

Dual theorem (dual problem): A term- used in linear programming to indicate a reverse problem to its counterpart, often styled the primal problem.

Dummy observation: In variance analysis with disproportionate subclass numbers it has been sometimes possible to get good approximations to the correct analysis by adding 'dummy' observations. These observations are usually inserted with values equa' to the means of the particular cells concerned.

Dummy variable: It is a quantity which is written in a mathematical expression in the form of a variable although it represents a constant; for example, in a regression equation

$$y = \beta_0 + \beta_1 X_1 + \beta_2 X_2 + \beta_p X_p$$

it has been more convenient to attach to the coefficient β_0 a dummy X_0 which has been always unity so that the expression may be written

$$y = \sum_{i=0}^{p} \beta_i Xi$$

Due-Trio test: It refers to a test in which three objects, two of which are alike are presented to a judge who tries to select the dissimilar object after being given the identity of two of them.

Duplicate sample: A sample which is collected concurrently, *i.e.,* in the course of the same sample survey under comparable conditions, with a first sample. It acts in much the same way as a replication except that in some cases the only thing which can get replicated has been the act of taking a second independent sample.

Duplicated sample: It is a sample which has been taken up twice for enquiry by two different parties of investigators. Two interpenetrating samples assigned to two investigating parties sometimes get so arranged, for purposes of control of field operations, as to have some common sample units these common units would constitute a duplicated (sub.) sample.

Durbin-Watson statistic: One of the assumptions of the ordinary least-squares method of estimating regression coefficients is that the disturbance, (stochastic) terms are uncorrtelated with one another.

A test for autocorrelation has been devised by Durbin and Watson. If $e_1, e_2, \ldots e_n$ are the unexplained residual from the fitted regression equation Me Durbin-Watson statistic is

$$d = \frac{\sum_{t=2}^{n}}{\sum_{t=1}^{n}}$$

and its value lies between 0 and 4. If the model stochastic terms are uncorrelated (at tag 1) d will take values near 2 smaller values indicate positive autocorrelation and higher values negative.

Dvoretsky's stochastic approximation theorem: It is a general theorem on the convergence of transformations with superimposed random errors. It is applicable to the stochastic approximation procedures of Robbins Munro and Kiefer-Wolfowitz.

Dynamic model: In econometrics a model is considered to be dynamic if it is having either or both of these properties: (1) at least one variable occurs in the structural equations with values taken at different points of time or in the form of time derivatives, etc.; (2) at least one equation contains a function of time.

Dynamic programming: It is a method which is used to secure a sequence of consistently optimum solutions to multi-stage decision problems, sequential decision considering in fact.

E

Econometric model: Refers to a set of one or more structural equations whose form (i.e., which variables depend on which) gets derived from economic theory and in which any unknown parameters are estimated statistically. There have been normally as many structural equations in the model as there are endogenous variables that the model seeks to explain.

Each equation relates one endogenous variable to other endogenous and exogenous variables. A simple example has been the two equation model of the market for an industrial product consisting of supply and demand questions. The endogenous variables in this case would be price and quantity, and exogenous factors might include advertising expenditure (affecting demand) and research and development (*r* & *d*) expenditure (affecting supply). Thus we might have

Quantity = $a+b$ × price + cX ($r\&d$)+ $u_1 \rightarrow$ supply equation

Quantity = d + eX price + fX advertising + $u_2 \rightarrow$ demand equation

These may be written as:

$$y = a + bx_1+cx_1+u_1$$

$$y = d + ex_1+fx_3+u_2$$

where x_1 denotes price, x_2 denotes research and development, x_3 represents advertising and where a, b, c, d, e, f and coefficients to be estimated and u_1, u_2 are unexplained residual terms. Given values for the exogenous variables which can be fixed by the firms, we can solve the equations simultaneously to find values for price and quantity to be sold.

Edgeworths' series: It is a method for representing certain skew frequency distributions and is based upon an expansion in terms of the Normal distribution and its derivatives. It is very similar to the Gram-Charher series.

If the cumulants of the distribution have been represented

by k_1, k_2, etc., the Normal function $\frac{1}{\sqrt{(2\pi)'}} \frac{1}{2^2} 2$ by $a(x)$ and the process of differentiating $\frac{d}{d_x}$ by D, the Edgeworth expansion of a frequency function $f(x)$ with zero mean and unit variance could-be but as follows.

$$f(x) = \exp\left\{\sum_{j=3}^{\infty} \frac{(-D)^j}{j!} k_j\right\} \quad a(x)$$

where the exponential has been expanded as a power series before the operations with D have been carried out.

Effective range: A group of observations may have a limited number of outlying observations at either or both ends of the range. It is possible to obtain a very rough measure of dispersion by taking the 'effective' range, i.e.. the range after these outlying values have been removed. The removal may have to be a matter of subjective judgment and inferences based or effective range have been somewhat doubtful value; in fact the term itself has been not a good one.

Efficiency equivalence: It refers to a broad extension of the concept of asymptotic relative efficiency to the case of regression parameters in second order processes.

Efficiency factor: The efficiency factor or an experimental design may be expressed in terms of a ratio of error variances, or, from a slightly different viewpoint, in terms of the size or number of replications required to attain a given precision.

Efficiency of an estimator: For an estimator which is asymptotically (for large samples) normally distributed and consistent, efficiency is defined as the ratio of sample sizes required to make its sampling variance equal to that of the efficient estimator as the sample size increases. Since sampling variances will depend on sample sizes this is just a way of assessing just how good a particular estimator is relative to the best one available for large samples.

Efficiency, statistical: The extent to which a statistic, or value obtained for a given sample, provides the maximum possible amount of information about one or more characteristics of a population.

Efficient estimators: most common estimators have asymptotically (for large sample sizes) normal sampling distribution. Consistent estimators wilt also be asymptotically unbiased. Among such estimators the one with the smallest variance in large sample is called an efficient estimator. This acts as a standard against which the usefulness of other estimators may be judged.

Eisenhart models: In variance analysis it has been customary to draw a distinction between type 1, in which the classification variables have been fixed, and type 2 in which they have been themselves random variables. The distinction has been pointed out by Eisenhart whose name has been frequently attached to the models, especially in balanced incomplete block experiments.

Element: A member of a group, set or collection of items which cannot be further subdivided into constituents that could be regarded as members of a collection. Hence the mathematical and statistical usage of the word is analogous to its usage in the physical sciences.

Elementary unit: Refers to one of the individuals which, in the aggregate, compose a population: the smallest unit providing information which, by suitable aggregation, gives rise to the population property under investigation.

Element sampling: Refers to a sampling scheme in which the sampling unit contains only one element. This is in contrast to cluster sampling, in which groups of elements are selected together.

Ellipticity: A term which is sometimes used to describe a bivariate normal distribution in which the variances have been not equal, leading to the idea of a 'test of ellipticity', i.e., a test of the inequality of the variances.

Elliptically symmetric distributions: This name is assigned to distributions whose density functions depend only on quadratic functions of the variables so that contours of constant density are elliptical. They are including spherically symmetric distributions as a special case.

Empirical: Derived from or relating to experiment and observation, rather than theory. An example of an empirical law has been the Rank Size Rule, which is used in geography. This states that if P_1 is the population of a country's largest city (or town), then P_1/n is that of its nth city. There has been no theoretical basis for this rule; it has just been found to work in many cases, as in the case of the USSR.

Rank	*City*	*Actual size (X 1000)*	*Predicted size (X 1000) (Bank Size Rule)*
1	Moscow	8,011	8,011
2	Leningrad	4,588	4,006
3	Kiev	20144	2,670
4	Tashkent	1,779	2,003
5	Baku	1,550	1,602
6	Kharkov	1,444	1,335

Empirical distribution function: Given an ordered sample of independent observations.

$$x_{(1)} < x_{(2)} < x_{(3)} < ... > x_{(n)},$$

the function $S_n(x)$ defined as:

$$S_n(x) = \begin{cases} 0, x < x_{(1)} \\ \frac{k}{n}, x_{(1)} < x < x_{(k+1)} \\ 1, x_{(n)} < x \end{cases}$$

is termed as an empirical distribution function.

Empirical probit: The analysis of experimental results concerned with the relationship between levels of a stimulus and the responses thereto has been generally carried out in terms of the percentage of test subjects reacting to the stimulus. The analysis often gets facilitated by transforming the percentages into probits. The probits corresponding to these observed percentage are termed as empirical probits, in contradistinction to expected probits and working probits, which get related to a fitted regression line.

Employment function: An equation or mathematical model expressing the average labour force employed as a function of the scale of production, past experience of the product or products being produced, and the average size or weight of the product. In effect it consists of a form of the production function, which expresses output in terms of quantity as a function of the average labour force, past experience and the average size of the product.

Empty cell test: It is a rather inefficient class of distribution-free tests which is characterised by the test statistic being the number of cells in a uni-or multivariate classification into which none of the sample members fall.

End corrections: An expression which is used in several statistical, contexts, to denote, corrections made to extreme values. For example, in taking a systematic sample from material which varies continuously it may be possible to got some increase in accuracy of estimation by using corrections to the term at each, end of the sample. Again, in fitting a curve to a time series, corrections could be made to end values to improve the fit. The expression is also found in the calculation of serial correlations where certain alterations become necessary to formulae correlating to infinite series to allow for the fact that the observations are having finite length.

Endogenous: In multivariate analysis the variable can be divided into two classes endogenous and exogenous. Those endogenous to the system are those which are influenced by other variables within the system.

Endogenous variable: This is a variable in an econometric model that is explained by the model. Thus it plays the same role as the dependent variable does in a regression equation. The model itself is a simplified representation of a complicated system in which many endogenous variables are being simultaneously determined. The driving force of the system is provided by the exogenous variables those whose values are fixed without reference to the system under discussion.

Endogenous variate: The statistical representation and analysis of multivariate systems all generally involving a primary division of variates into those which are endogenous and those which are exogenous. Endogenous variates have been those which form an inherent part of the system, as for instance price and demand in an economic model.

Ensemble: Refers to the infinite set of realisations of a single stochastic process. The concept has been frequently extended to apply to a sample set of realisations.

Ensemble average: Given a number of independent sample realisations from the same generating process, it is possible to estimate the mean of the process by the overall mean of the samples. The estimator is termed as an 'ensemble average'. If the process has been stationary, its mean can be estimated by a simple average over time instead of by the 'ensemble average.

Entry plot: A plot through which a serial cluster gets entered or determined.

Enumeration data: Refers to data which is consisting of numbers of individuals in various categories.

Example: a flower is having three varieties, red, white and pink, 400 flowers are collected and found to be:

Red	89
Pink	201
White	110

There have been thus three categories, red, pink and white the data consist of the number of flowers in each.

Envelope power function: A power function which is based on a critical region which., itself is obtained as the envelope of a number of critical regions.

Epsem sampling: The word 'epsem' denotes an Equal Probability of Selection Method. Epsem sampling therefore includes any sampling scheme in which every element in the population has an equal probability (chance) of selection. The big advantage of epsem is that it leads to self-weighting samples, in which there is no need to give different weights to different elements in order to obtain good estimates of the population parameters.

Equal probability of selection method: It is any method of sampling in which the population elements are having an equal probability of selection. 'EPSEM' sampling could result from either equal probability selection throughout, or, from variable probabilities that compensate each other in multi-stage selection.

Equal spacings test: It is a non-parametric test of uniformity for circular distributions which is invariant under rotation.

Equal-trails test: Refers to a symmetrical double-tailed test.

Equally correlated distribution: A multivariate distribution has been equally correlated if the correlation between each pair of variates has been equal to ρ.

Equilibrium: In queueing theory, systems in 'statistical equilibrium' have been those in which the number of customers or items waiting in the queue oscillates in such a way that mean and distribution are constant over a long period.

Equilibrium distribution: A physical, economic or social system which gets settled down to a stable statistical behaviour is having a stationary, or equilibrium, distribution which specifies the limiting proportions spent in the designated states of the system.

Equiprobable: (Of values) of equal probability. The possible scores it a die is thrown, 1, 2, 3. 4, 5, and 6, are equiprobable, each is having a probability of occurring of 1/6. The scores 0 to 36 on a fair French roulette wheel, have been also equiprobable, at 1/37 each.

Equitable game: It is an alternative name for 'fair' game.

Equivalent samples: Any two samples which are selected by the same sampling scheme and contain the same set of d ($\leq n$) population members, yield the same information about the paramet:rs, irrespective of whether these d members appear with different frequencies in the two samples.

Ergodicity: Generally, this term denotes a property of certain systems which develop through time according to probabilistic laws. Under certain circumstances a system will tend in probability to a limiting from which has been independent of the initial position from which it started. It is the ergodic property.

Erratic movements: In a time series these are irregular movements as opposed to trend or cyclical components. They include the ordinary random fluctuations that continually occur but exclude longer departures than can be traced to specific and isolated incidents such as wars and strikes; these latter are referred to as episodic movements.

Error: In general, refers to mistake or error in the colloquial sense. There may, for; example, be a gross error or avoidable mistake; an error of reference, when data concerning one phenomenon get attributed to another; copying errors; an error of interpretation.

Error, alpha: The rejection of a true null hypothesis. Since null hypotheses are accepted or rejected on the basis of statistical probability, the possibility exists that a true null hypothesis may be erroneously rejected.

Error band: In estimation or prediction the estimated or predicted value gets bracketed by a range of values determined by standard errors, confidence intervals or similar methods within which the value may be regarded to lie with a certain probability. This is termed as the error band.

Error beta: The retaining of a false null hypothesis. Since null hypotheses are accepted or rejected on the basis of tests of statistical probability, the possibility exists that a false null hypothesis may be accepted. The degree of danger of accepting a false null hypothesis depends upon the level of significance used in testing the hypothesis: the lower the level of significance used, the greater the probability of making a beta error. Beta error is sometimes referred to as type II error. See error, alpha, hypothesis, null; probability; significance, statistical

Error, chance: In statistics, one of a number of errors that arise from chance and tend to compensate for one another and cancel each other out Chance errors, sometimes referred to as random or accidental errors, are usually contrasted with constant error or bias.

Error curve: The normal distribution curve.

Error in equations: An equation in variables or variates may be inexact either because the equation has been not a complete representation of the situation, as in a demand supply equation which does not include other factors such as income or employment, or because it gets disturbed by extraneous sources of variation as in an autoregression equation.

Error mean-square: It is equal to the residual or error sum of squares divided by the number of degrees of freedom on which the sum gets based. It gives an estimator of the residual or error variance.

Error of estimation: In general, refers to the difference between an estimated value and the true value.

Error of first kind: If, because of a statistical test, a statistical hypothesis gets rejected when it has to be accepted, i.e., when it has been true, then an error has been committed. This class of error is termed an error of the first kind and is fundamental to the theory of testing statistical hypotheses associated with the names of Neyman and E.S. Pearson.

Error of observation: An error which arises from imperfections in the method of observing a quantity, whether due to instrumental or to human factors.

Error sum of square: The sum of squares of residuals between observations and estimates using a particular multiple regression model. In analysis of variance it is assumed that observations result (1) from the effect of a number of regressors whose variances can be estimated and (2) from a random component, i.e.,

$$y = \sum_{i=1}^{p} \beta_i x_i + \varepsilon$$

where a linear model is used.

Error variance: Refers to the variance of an error component. Thus, if the generating model of a set of data is having certain systematic components together with a stochastic component, the variance of the latter is the error variance.

Esseen's Lemma: This result was due to Esseen. This formula gives an estimate of the closeness of two distribution functions F and G in terms of the closeness of the characteristic functions on an arbitrary interval –T, T. It states that if F and, G have been distribution functions with respective characteristic functions f and g, and if G is having first derivative G, then

$$\text{Sub } |F(x)\text{-}G(x)| \leq \frac{1}{\pi}\int_{-T}^{T}\left|\frac{f(t)-g(t)}{\pi t}\right|dt + \frac{24}{\pi T}\underset{x}{Sub}\,-G'(x)|.$$

Estimate: 1. Refers to an approximate calculation. 2. Refers to an approximate idea of something (e.g., size, cost). 3. To make an approximate calculation, or to form an approximate idea of something (e.g., size, cost).

Estimating equation: A mathematical formula that relates the unknown variable to the known variables in regression analysis.

Estimator: A random variable, formula or procedure for estimating the value of a parent population parameter from sample data. Its value depends on the particular sample involved. An estimator should ideally be consistent, unbiased and efficient.. Estimated values of population parameters are distinguished from their true values by the convention of using the symbol

$\hat{\sigma}$ .Thus σ = true parent population standard deviation while

$\hat{\sigma}$ = estimated (from a sample) parent population standard deviation.

Event: In probability this term is employed to mean any one of a set of probable outcomes or happenings, which in simple illustrations of probability are always mutually exclusive. For example, it a die is cast, the outcome '6' is an event, -which excludes the outcomes '1', '2', '3', '4' and '5'.

Evolutionary process: Refers to any non-stationary process. The probability distributions associated with the process have been not independent of the time.

Exact' statistical method: It is used in interval estimation to imply that the probability distributions involved have been completely known and that probability levels quoted have been exactly achieved.

Exceedances tests: Refers to a class of distribution free tests of identity of two populations which are characterised by having a test statistic that is the number of observations in a simple random sample from one population that exceed a particular order statistic in a simple random sample from the other population. See exceedances, distribution of.

Exclusive: (Of two events) such that both events cannot take place. Thus events A and B are said to be exclusive if B cannot take place when A does, and *vice versa*. It has been however, possible that neither A nor B takes place.

Example:

A: getting a 6 when throwing a die

B: getting a 4 on the same throw of the die

When three or more events have been such that no two of them can take place at the same time, they are considered to be *mutually exclusive*. Compare disjoint.

Exhaustive: 1. (Of two events) such that at least on must occur. Thus events P and Q are said to be exhaustive if either P alone, or Q alone, or both P and Q together, must take place it is' not possible for neither to take place. Exhaustive events cover all possibilities.

Example:

P: An athlete does a track event

Q: An athlete does a field event

As all athletics events have been track, or field, or both

(e.g., the decathlon), P and Q are exhaustive events. 2. (of a greater number of events) such that they cover all possibilities between them, they are comprehensive in their scope.

Exhaustive classification: A classification or categorization of the variate values which includes all possible values. If, in addition each possible value can belong to one and only one category the classification is said to be exhaustive and mutually exclusive.

Exhaustive sampling: This term is occasionally encountered. This term indicates that, in order to achieve the sample size necessary for the required precision, the sampling has included all of the population under study. That is to say that a complete enumeration rather than a sample is taken.

Exogenous: In multivariate analysis the variables can be divided into two classes—endogenous and exogenous. Those exogenous to a system are those determined outside the system and not influenced by other variables in the system (or such determination is not of interest in the model).

Exogenous variable: In an econometric model an exogenous variable is one that is determined outside the model and is not itself jointly dependent with the endogenous Variables explained by the model.

Exogenous variate: This term is the opposite to an endogenous variate.

Expectation: Refers to the expected or mean value of a-random variable, or of a function of the variable. Symbol: E ( ). For a discrete variable, X, taking values $x_1, x_2, \ldots x_n$ with probabilities $p(x_1)$, $p(x_2)$, ..., $p(x_n)$, the expectation of X, E (X) or $\overline{X}$ would be given as follows

$$\sum_{i=1}^{i=n} x_i p(x_i)$$

and for a continuous variable

$$\int xf\ (x)dx$$

where $f(x)$ denotes the probability density function, of X.

The 'expected' value is not necessarily the most frequently occurring value or even a possible value; e.g., if a variate cat take each of the values 0 and 1 with probability ½ and no other value is possible, the expected value is ½.

Expected frequencies: The frequencies we would expected to see in a contingency table or frequency distribution if the null hypothesis is true.

Expected value/expectation: The mean value of an estimator or function in repeated sampling is its expected value. Thus, if an estimator could take the value 3, 7, 12, 2, 15, 9 each with probability 1/6, the expected value of the estimator is 1/6[3+7+12+2+15+9] = 48/6 = 8. Thus the expected value need not be a value which itself occurs at all. For example, with a fair die (each side having equal probability of occurring in a single toss) the expected value is 1/6[1+2+3+4+5+6]x3½, which can never occur. The symbols E[ ], ε[ ]and EV[ ] are commonly used to denote expected value.

Expected value criterion: A criterion requiring the decision maker to calculate the expected value for each decision alternative (the sum of the weighted payoffs for that alternative in which the weights are the probability values assigned by the decision maker to the states of nature than can happen).

Expected value of a random variable: The sum of the products of each value of the random variable with that value's probability of occurrence.

Experiment: Any process which results in the collection of data. (This meaning for the word has been somewhat different from that usually given to it in other scientific disciplines.)

Examples: Of experiments in a statistical sense are carrying out a survey; measuring the daily rainfall; recording the longevity of animals of a particular type.

Experimental design: Refers to the way in which a statistical experiment is planned. The example which follows illustrates the use of a number of different designs, namely: completely randomized design, randomized blocks design, Latin square, and Graeco-Latin square.

Experimental error: Refers to the difference between the value of a quantity found in an experiment and that which should have been found had the experiment been conducted perfectly. Experimental error has been due to experimental technique, and does not include mistakes in mathematics or experimental design.

More specifically, the term experimental error is used to denote the essential probabilistic variation to be expected under repetition of the experiment, not actual mistakes in design or avoidable imperfections in technique. It has been the aim of good experimental design to provide valid measures of the experimental error in the more restricted sense.

Experimental or empirical probability: Refers to a probability which is determined by experiment or observation rather than from some underlying theory. The figures used by life insurance companies in calculating the risks for particular individuals are included into this category. On the other hand, a skilled bridge player decides his strategy on the basis of theoretical probabilities.

Explosive process: It is a stochastic process which is having no bound to the expectation of the mean square. A process the values of which may increase without limit (in absolute magnitude, so that oscillations are possible) as time goes on the term is not to be recommended.

Exponent: The power to which a number is raised. Hence if a number is squared the exponent is 2. In the case $10^{8.8}$ the exponent is 8.8.

Exponential: (Of a value or function) raised to the power of. Thus the function $F(x) = a^x$ has been an exponential function; x has been called the exponent.

The term exponential has been most commonly used when the exponent has been the power of e, the base of natural logarithms. In this case the exponential of x has been e^x which has been sometimes written as $\exp(x)$.

Exponential distribution: It is a continuous distribution with probability density function given by:

$$f(x) = \lambda e^{-\lambda x}$$

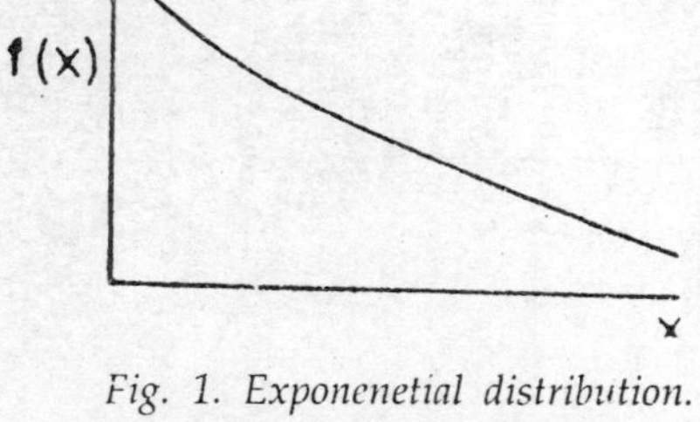

Fig. 1. Exponenetial distribution.
For this distribution:

$$\text{Mean} = \frac{1}{\lambda} \quad \text{Variance} = \frac{1}{\lambda^2}$$

The exponential distribution has been sometimes used in connection with the length of life of certain materials.

Exponential regression: A term which is used to denote a relationship of the form.

$$y_i = a + \beta e^{\lambda x t} + \varepsilon$$

Where ε denotes a random residual. The term 'asymptotic regression' would be preferable.

Ex Post: 'From afterwards.' The term generally means a retrospective view of an event or set of events. Actual observed results may be compared with those predicted by a given model, and the resultant comparison ex post. Any modification of the model or hypothesis is also termed ex post. In financial statistics the term is sometimes employed for modified budget estimates, taking into account exceptional events which have occurred since the original ex ante estimates were made.

Extensive sampling: A term which is used to denote sampling where the subject matter, or geographical coverage of a Sample has been diffuse or widespread as opposed to intensive, where it gets narrowed to a small field. Extensive sampling. May refer either to a case where a wide variety of topics have been covered superficially, rather than a few topics in detail or a large area has been surveyed broadly, rather than a small area studied in detail. The term might also be used with reference to time, that is to say, of sampling covering a long period.

Extremal quotient: The relative difference between the largest and smallest values in a sample, as the range is the absolute difference.

External variance: An obsolete term which was introduced by Schultz. This term is used to denote the variance of predictions of the future movements of a particular time series which is based upon a given form for its-trend. The components contributing to the variance have been errors in the estimation of the parameters, not those arising from an incorrect choice of trend line or from superposed random variation.

Extrapolation: The extension of a graph beyond the upper or lower limit of the observed range of the basic statistics, or the calculation from a regression equation of an estimated value or values of the dependent variable corresponding to a value or values of one or more of the independent variables outside the observed range. It is a process which theoretical statisticians frown upon, and indeed, it needs using with case. The further the graph is extended beyond the observed range the more dangerous the process becomes; the same applies to extrapolated values calculated from a regression equation.

Extremal statistic: Refers to any function which depends upon observations at both extremes of the sample. Examples of external statistic include the range and the extremal quotient.

Extreme mean: Refers to one of a set of mean values, e g., in the analysis of variance which lies at the extreme of those values, i.e., the largest or the smallest.

F

Factor: This word finds use in statistical contexts in many senses:

(a) in the ordinary mathematical sense, e.g., it denotes a factor of an algebraic expression.

(b) it is also used to denote a quantity under examination in an experiment as a possible cause of variation, e g., in *a* 'factorial' experiment;

(c) in multivariate analysis, it is used to denote a function of the observed variates, usually linear, which may be regarded as part of those variates and hence as a 'factor' of the variation;

(d) it is used to denote a constituent item in an average or index number.

Factor analysis: A statistical procedure used to determine the basic irreducible variables (factors) underlying a large number of interrelated variables. When measurements of a large number of variables have been obtained factor analysis may be used to reduce them or a smaller number of basic types or factors. Factor analysis involves the analysis in essence involves the analysis of the correlations among the variables.

Factorial: This term refers to the product of all the positive integers from I up to and including a given integer. Symbol *n*! where *n* is the given integer. Thus 5 factorial, written 5!; meanes

$5 \times 4 \times 3 \times 2 \times 1$ (= 120)

Factorials find use in working out probabilities in the poisson distribution and binomial distribution, in permutations and combinations, etc.

Factorials very quickly become large numbers. By convention, 0! is given the value

1 ! = 1

2 ! = 2

$3! = 6$

$4! = 24$

$5! = 120$

$6! = 720$

$7! = 5{,}040$

$8! = 40{,}320$

$9! = 362{,}880$

$10! = 3{,}628{,}800$

For large values of n, $n!$ may be given approximately by

$$n! \simeq \sqrt{2\pi n}\, n^n e^{-n}, \text{ or}$$

$$n = ! \simeq \sqrt{2\pi}\, n^{(n+½)} e^{-n}$$

which can be rewritten as *Stirling's approximation to n!* $\log_{10} n! = .39909 + (n + ½) \log_{10} n - .4342945\ n$

The concept is important in such cases as :

1. calculating probability in sampling without replacement
2. calculating expected values in the poisson distribution;
3. computing probability in fisher's exact probability test. The term should not be confused with factor.

Factorial experiment: An experiment which has been designed to examine the effect of one or more factors, each factor being applied at two levels at least so that differential effects might be observed. The term finds frequent use in a slightly narrower sense, as describing an experiment investigating all possible treatment combinations which may be formed from the factors under investigation. The 'level' of a factor denotes the intensity with which it is brought to bear. It could be measured quantitatively, as when fertilizer is applied to plots in a given weight per unit area, or qualitatively, as when patients are considered at two levels inoculated' and not inoculated, gives rise to calculations for analysis of variance.

Factorial sum: Refers to the sums- which are entering into the calculation of factorial moments. If the frequency of a variate value $r(r = 0, 1, .., k)$ has been f_r the factorial sum of order would be.

$$\sum_{r=0}^{k} \{f_r r\,(r-1)\ldots(r-j+1)\}$$

Factor rotation: The final stage of factor analysis has been in geometric terms, a rotation of the factor axes so as to achieve some correspondence between the numerically derived factors and explanatory entities; that is to say to seek a structure.

Analytic methods could be divided broadly into those dealing with orthogonal factors, where the loadings get uncorrected and oblique factors involving correlated loadings. One group of methods pursues the maximisation of the scatter among factor loadings for example, quartimax and varimax for orthogonal rotation and oblimax for oblique rotation. An alternative approach, using the minimisation of cross products, is called qartimin and covarimin, which latter method is the inverse of varimax.

Fair game: In the theory of games, a game consisting of sequence of trials has been deemed to be a 'fair' game if the cost of each trial is equal to the expected value of the gain from each trial. A 'fair' game in this sense may not be fair as between a pair of adversaries with unequal resources: it is well known that at a 'fair' game the player with the larger sum to stake has the better chance of ruining his opponent.

Fertility rate: Refers to the number of live births in a unit period which is expressed, as a proportion of potentially fertile women in the population concerned. 'Potentially fertile' is generally defined by reference to age, e.g., be taking women from 15-50 years old. This crude or general fertility rate may be refined or standardised in respect of age, social class, etc.

Fertility rate is usually expressed per thousand per year.

Fiducial distribution: Refers to a distribution of a parameter needed for fiducial inference about that parameter. It is not a probability distribution in the customary sense, but gets derived from the distribution of estimators containing all the relevant information in the sample. In earlier literature it was sometimes referred to as a fiducial probability distribution.

Fiducial inference: A type of statistical inference which is based on fiducial distribution. It was introduced by R.A. Fisher. The object of the inference has been to made probabilistic statements about the values of unknown parameters and to that extent it resembles the theory of confide intervals.

Fiducial limit: Sometimes called fiduciary limit or confidence limits; the limits within which a derived figure such as an average of a standard deviation is true of the universe.

The fiducial probability that the true average of the universe falls within the fiducial limit set by the standard deviation of the statistical series applied to the average of the series, is 68%. One and a half times the standard deviation has a fiducial probability of 87% and twice the standard deviation has one of 95%. If the average of a series is 156, and the standard is 20, then the fiducial probability is 68% that the true mean of universe falls within the fiducial limits of 156 ± 20 i.e., between 136 and 176.

Financial ratios: Ratios between particular groups of the assets or liabilities of an enterprise and corresponding totals of assets or liabilities; or between assets of liabilities and flows like turnover or re venue.

Finite Markov chain: A Markov chain fix is termed as a finite chain with A: states when the number of possible values of the random variables $\{x_n\}$ is finite and equal to k.

Finite population: Refers to a population of individuals which have been finite in number.

Finite population correction: A correction factor which is applied to the expression for the variance of the means of samples drawn from an infinite parent population, to allow for it being of finite size.

Finite sampling correction: Refers to as alternative term for the finite multiplier. It has been thought as objectionable in the grounds that it has been not a correction to an inexact result,

First limit theorem: If a sequence of distribution functions is tending to a single distribution function F then the corresponding characteristic functions would tend uniformly in any finite interval to the characteristic function of F. This is generally termed as the First Limit Theorem. Although it is known to earlier writers, it seems to have been proved rigorously for the first time independently by p.

Fiscal federalism: A system of taxation and public expenditure in which revenue raising powers and control over expenditure are vested in various levels of government within a nation, ranging from the national government to the smallest unit of local government.

Fisher's 'B' distribution: Refers to the distribution of the square root of a non-central chisquared variate. It has been a hypergeometric function and was derived by Fisher in a study of the multiple correlation coefficient.

Fisher's ideal index number: Refers to the geometric mean of Laspeyres and Paasche's index numbers. Laspeyres and Paasche are biased, if at all in opposite directions; where the index is one of prices, the former usually upwards and the latter downwards, Professor Irving Fisher, the American economist and author of the standard work on index numbers. *The Making of Index Numbers*, suggested that a cross of Laspeyres and Paasche would therefore give the ideal index number. The subject is more fully discussed under index numbers.

Fisher model: A termhich was proposed by Ogawa to denote the general class of experiment designs without technical errors but based upon Fisher's randomisation procedure.

Fisher's distribution: Refers to the distribution of the ratio of the variances of two independent samples from a Normal population or, a little more generally, to the ratio of two independent quantities, each of which gets distributed as z^2.

Fisher exact probability test: A nonparametric statistical technique used for analyzing discreate data from two samples that are small in size. The Fisher test is used to determined whether there is a significant difference between two samples in the proportion of cases that fall in each of two mutually exclusive categories. The Fisher test is used in preference to the chisquare test when the number of cases is small and the data are in a fourfold (2 × 2) table.

It is a method of testing association in (2 × 2) contingency table where the total number of observations is less than fifty, and where in consequency, Yates modification of the chisquared test is unreliable.

Fisher's z-transformation: Refers to a transformation of the correlation coefficient according to the formula.

$$z = \tan h^{-1}_{r}$$

The distribution of z for samples from a bivariate Normal population approaches normality of form much more rapidly than does that of r; and its variance, unlike that of r, has been little affected by the population correlation coefficient even for samples of moderate size. It has been therefore useful in many problems involving estimation and tests of significance.

Fisher's z transformation has been used:

(a) In testing the hypothesis that the correlation coefficient of the parent population of a bivariate distribution has a particular value :

(b) In up confidence limits for an estimated value of the parent population s correlation coefficient.

When a correlation coefficient is calculated, the answer based on a same. If, for example, 20 Siames cats were used in a dietary experiment, the correlation coefficient between their masses (in kg) and daily food intake (in calories) could be calculated, giving a value of, say, 0.6. The 20 cats are a sample from a very large parent population, all Siames cats. Thus the value of 0.6 is only an estimate of the parent population correlation coefficient, based on a sample of size 20.

When significance level tables has been used for interpreting correlation coefficients, the null hypothesis being tested is: There has been no correlation. The sample has been drawn from a parent population correlation coefficient 0.

The alternative hypothesis has been that the parent correlation coefficient is not zero. Thus a two tail test is involved. In order to test this hypothesis, it should be necessary to know the distribution of correlation coefficients for samples of size 20.

Fixed base index Refers to an index number for which the base period for the calculations has been selected and remains unchanged during the lifetime of the index. It has been in contradistinction to a chain base index.

Fixed-base index number A series of index numbers with a common base (=100). The base usually covers one of the periods, times or places within the series, not necessarily the first in a time series. It serves as a common standard of comparison. The weighting system does not necessarily follow the base.

The pattern of weights applies to the base period or place when Laspeyres's index is used, and to the given period or place when Paasche's index is used. A chain-base index applicable only to time series, uses the weights either of the given period or of the preceding period, the base always being the preceding period.

Fixed effects (Constants) model: Refers to an alternative designation for model I (or first kind) in analysis of variance.

Fixed sample: If a survey is repeated on several occasions, but observations are taken on the same sample instead of a new sample for each occasion, the sample is considered to be fixed.

Fluctuation: Refers to a movement up or down between consecutive items of a series of numbers or numerical observations. In a different sense the variation of a statistic from sample to sample is also termed as a sampling fluctuation.

Forecasting 'Forecasting' and 'prediction' have been generally employed synonymously in the customary sense of assessing the magnitude which a quantity will assume at some future point of time: as distinct from 'estimation' which tries to assess the magnitude of an already existent quantity. For example the final yield of a crop has been 'forecast' during the growing period but 'estimated' at harvest.

The errors of estimation involved in prediction from a regression equation have been sometimes referred to as 'forecasting errors' but this expression has been better avoided in such a restricted sense. Likewise terms such as 'index numbers of forecasting efficiency' in the sense of residual error variances in regression analysis have to be avoided.

Foreign investment: This usually refers to any investment in another country which is carried out by private companies or individuals as opposed to government aid. There are arguments for and against foreign investment in underdeveloped countries: those against, say that it constitutes economic colonialism and is an attempt to retain control of ex-colonies.

Fourfold table: A statistical tables with four cells, or places for the entry of data. A fourfold table is formed by the cross tabulation of two variables each of which has just two categories, so that there are four possible combinations of categories. For example, if the variable sex, with the two categories men and women, and the variable schooling, divided into just two categories—graduated from high school and did not graduate from high school were cross—tabulated, would be four possible combinations of categories graduated from high school men who did not graduated from high school women who graduated from high women who did not graduate from high school.

Fourier analysis: Refers to the theory of representing f of a variable t as the sum of a series of sine and cosine terms of type a_j cos $(2\pi/\lambda_\varphi)$, j = 0, 1. The l 's has been not necessarily commensurable and hence the analysis has been more general than harmonic analysis which considers series of terms such as cos $(2\pi_j/\lambda)$ whereλ's has been some constant.

Frame: Refers to a representation of the items which are available to be chosen for a sample. This could be a telephone directory, a map an electrical register, a list of the cattle in a herd, etc.

Freehand method: Refers to a method of describing the relationship in a series of data, ordered in time or space, whereby the general trend has been estimated by drawing a line freehand through or near the series of plotted observations.

Frequency 1. The number of occurrences of a particular event ; or 2. a particular class or interval group ; or 3. particular values of a variate,

For example, if in a degree examination there are three distinction results, twelve passes with credit, thirty ordinary passes and two failures, then the frequency of distinctions in the examination is three, etc,

Example: in 20 shots at a

target, a marksman makes the following scores

4 3 2 4 5 2 4 2 5 1

0 5 1 5 2 5 5 0 1 4

The frequencies of the different scores are thus:

Score	0	1	2	3	4	5
Frequency	2	3	4	1	4	6

Frequency curve: Refers to the graphical representation of a continuous frequency distribution, the variate being the abscissa and the frequency the ordinate. The frequency curve could be viewed as the limiting form of the frequency polygon as the number of observations tends to become infinitely large, and the class intervals indefinitely small. Thus frequency curves are used where class intervals be more very small and numbers of observations are very large. The ordinate axis is used to measure frequency values and the abscissa is employed to measure variate values.

Frequency distribution: A classification of data showing the number of occurrences of each class interval subdivision of quantitative variable or category of a variable, In the case of a quantitative variable the class intervals are arranged magnitude. A frequency distribution provides a order of basic ordering of data, and is usually the first step, in statistical analysis.

Frequency distribution bimodal: A frequency distribution having two modes or peaks. In a bimodal frequency distribution there are two separated (noncontiguous) class intervals have the largest number of cases in the distribution of a variable.

Frequency distribution cumulative: A frequency distribution that presents for each class interval (subdivision of a quantitative variable) the total number of cases in that interval and below (or above) it. Thus the number of cases given for each class interval includes the cases that fall in that interval and in all perceiving intervals.

Frequency distribution multimodal: A frequency distribution with three or more modes.

Frequency distribution symmetrical: A frequency distribution in which the distribution of cases is identification both sides of the mode (most frequent value, or clast interval with the greatest a number of cases). In a symmetrical frequency distribution the mode, the median and the arithmetic mean coincide.

Frequency distribution unimodal: A frequency distribution that has one model or peak of frequencies.

Frequency distribution curve: The graphic presentation of a distribution The frequencies are listed along the vertical axis and the class intervals or their midpoints are listed along the horizontal axis.

The frequency of each class interval is then plotted and as drawn connecting these points. A frequency distribution curve is often referred to as a frequency Polygon.

Frequency distribution curve, leptokurtic: A frequency distribution, curve having a pronounced peak or a higher degree of kurtosis than a normal frequency distribution curve.

Full Information method: In econometrics, it refers to a method of deriving estimates of parameters in a stochastic model which have been subject to all the a *priori* restrictions of that model.

Function: In statistics used in the mathematical sense that is to quote the Oxford Dictionary variable quantity in relation to other in terms of which it may by expressed or on which its value depends.

Mathematicians and mathematical statisticians use a form of shorthand, e.g.,

$$Y = f(x)$$

which simply means that Y is a function of X. In practical statistics an equation like this serves merely as a preliminary model: it saves writing out a verbal model. Before it can be applied in practice, it has to be converted to an equation which indicates the type of relationship that exists theoretically, or empirically, between the variable quantity and the other (s).

$$Y = aX + b$$

$$Y = bX^a$$

$$Y = aX^2 + bX + C$$

$$Y = ba^Z$$

Fundamental probability set: Refers to a set of objects or events which are basic to a probabilistic situation, in the sense that all other objects or events under consideration are derived from them by compounding. It is evident that all probabilities are expressible by the rules of addition, multiplication, etc., of probabilities in terms of the probabilities of the fundamental set. Failure to specify the fundamental set explicitly leads to confusion, occasionally to error, and in particular to a number of paradoxes. The fundamental probability set has been sometimes called the reference set.

G

Gabriel's test: Refers to a simultaneous comparison procedure for mean values in analysis of variance. It has been based upon the 'between-groups' sum of squares for the (2^k-k- 1) subsets- of two or more mean values of the k groups and has been tested against the fixed critical value of, the F-distribution with (k-1, n-k) degrees of freedom. The sample sizes (n_i) need not be equal.

Galton Ogive: Under certain conditions, e g., when a frequency function has been unimodal, the distribution curve resembles a letter 'S' the ogive form, a term which was borrowed by Galton from architecture and used particularly for the distribution curve of the Normal distribution.

Galton-McAllister distribution: Refers to a frequency distribution in which the logarithm of the variate, or some simple linear function of it Normally gets distributed. The distribution was suggested by McAllister. It is now more generally termed as the lognormal distribution.

Gambler's Ruin: Refers to the name which is assigned to one of the classical topics in probability theory. A game of chance can get related to a series of Bernoulli trials at which a gambler wins a certain predetermined sum of money for every success and losses a second sum of money for every failure. The play may take place until his initial capital gets exhausted and he gets ruined. The statistical problems involved get concerned with the probability of he ruin of a player, given the stakes, initial capital and chances of success, and with such matters as the distribution of the length of play.

There have been many variations to this classical problem, which gets closely associated with problems of the random walk; in particular, of sequential sampling.

Game theory: A game is a conflict situation and game theory is concerned with devising and giving effect to the strategies of such situations, namely the main features of a plan prepared before action commences and including an objective.

Other related terms have been as follows:

Pay-off—the value of the game to a player.

Zero-sum game—game in which all the pay-offs for all players total zero, i.e., what one player gains, other lose.

Non-zero-sum game—game in which all the pay-offs for all players do not total zero.

Finite game—game in which each player has a finite number of possible strategies.

Infinite game-game in which at least one player has an infinite number of strategies.

A 2, 3, 4 ... *n*-person game—game with 2, 3, 3...*n* players or conflicting sets of interests.

Game theory is an important branch of operational research and is related to decision theory and operational gaming. It is also linked closely with linear programming in that linear programmes can often be used to obtain the best strategies to use in a conflict.

Game theory deals with situations of conflict between two or more opposed sets of interests and is used to calculate the optimum strategy which maximizes the winnings or minimizes the losses of one or more of the 'players'. (The usual criterion is to minimize the maximum losses).

The working procedure consists of six steps as follows

(i) Formulate the problem.

(ii) Specify the pay-off criterion.

(iii) Set out the pay-offs in the form of a 'game matrix'.

(iv) Carry out the requisite arithmetic operations until the solution is obtained.

(v) Test the solution.

(vi) implement.

Gamma coefficients: Refers to the ratios, analogous to moment ratios or the beta coefficients, which have been based upon the cumulants:

$$\gamma_1 = k_3/k_2^{3/4},\ \gamma_2 = k_4/k_2^{2}$$

and, in general

$$\gamma_r = k_r/k_2^{r/2}$$

Gamma distribution: A particular kind of distribution, also known as type III in the Pearsonian system of frequency distributions. It is continuously distributed from zero of infinity and is also unimodel. The formula for the frequency distribution with probability density function f (x) is given by:

where a and λ have been parameters of the distribution, and λ (a) denotes the gamma function, given by:

$$f(x) = \frac{x^{a-1}e^{-\lambda x}}{\lambda^{-a}\,\lambda(a)}$$

where α and λ have been parameters of the distributions, and $\lambda(\alpha)$ denotes the gamma function, given by:

$$\lambda(\alpha) = \int_0^\infty x^{\alpha-1}\, e^{-x}\, dx \text{ (for a real and positive).}$$

If α has been a positive integer, $\lambda(a) = (\alpha\text{-}1)!$

The gamma distribution has

$$\text{Mean} = \frac{\alpha}{\lambda}$$

$$\text{Variance} = \frac{\alpha}{\lambda^2}$$

Gantt progress chart: Refers to on application of the bar chart due to Gantt, of use in industrial statistics. An actual Performance or output has been expressed as a percentage of a quota or planned performance per unit of time. Account May also be taken of the cumulative performance by plotting it, with the planned cumulative performance, as ordinate against time as abscissa.

Gauss distribution: Refers to an alternative name for the Normal distribution.

Gauss-Markov theorem: Refers to a fundamental theorem which deals with an unbiassed estimator of a population characteristic, and is based upon a linear combination of sample observations drawn from that population. The theorem has been to the general effect that an unbiassed linear estimator of a parameter has been 'best i.e., has minimum variance, when the estimator has been obtained by least squares. The theorem could be extended in many directions, e.g., by regarding the simultaneous estimation of several parameters or linear functions of them.

Gauss-poisson distribution: Refers to another name for the Poisson-Normal distribution.

Geary's ratio: As a test of normality the moment ratio for kurtosis is having the drawback that its sampling distribution has been skew even for quite high values of n, the sample size. In Order to overcome this, Geary pospulated a test in the form of a ratio

$$\frac{\text{mean deviation}}{\text{standard deviation}}$$

which, in samples from a Normal distribution, tends to $\sqrt{(2/\pi)}$ as n tends to infinity, The distribution of Geary's Ratio would tend to the Normal form fairly rapidly and, as a test, the ratio aims at detecting departures from mesokurtosis in the parent population.

General renewal process: Refers to a class of stochastic process in which the time to the first event is having distribution function $F_1(t)$, but thereafter the intervals between events have been independently and indentically distributed with distribution function $F(t)$. $F(t)$ has been termed as the 'renewal distribution' whilst $F_1(t)$ has been referred to as the distribution of the first renewal time.

The term renewal process has been used if $F_1(t) \equiv F(t)$.

Generalised binomial distribution: Refers to an alternative name for the Poisson binomial distribution.

Generalised bivariate exponential distribution: Refers to a generalisation of the bivariate form which is obtained directly from the multivariate exponential distribution. This generalisation, has been important in connection with the form of the underlying model used in life assessment problems.

Generalised distribution: This expression is found in two senses, (i) as a more complicated form similar to some known distribution; (ii) as obtained from a known distribution when its parameters are themselves random variables. The latter usage should get discarded in favour of compound distribution.

Generalised maximum likelihood estimator: Refers to an approach to parameter estimation which is involving the concept of asymptotic efficiency proposed by Weiss & Wolfowitz to remove some of the problems associated with classical maximum likelihood estimators.

Generalised multinomial distribution: Refers to an *n*-dimensional distribution of discrete variates which were proposed by Tullis which includes an additional parameter ρ as the correlation coefficient between pairs of variates with common mean and marginal distributions which have been multinomial.

Generalised normal distribution: A little-used alternative same for Kapteyn's univariate distribution which is not to be confused with the multivariate Normal distribution.

Generating function: Refers to a function of a parameter *t* which, when expanded as a power series in *t*, gives as the coefficients the values of some quantity of statistical interest such as the probability of events or the moments of a frequency distribution. The characteristics function has been an important case of a moment generating function, a phrase which has been often abbreviated to *m.g.f.* The theory of probability is making use of the generating function approach since the time of De Moivre in the first half of the 18th century.

Geometric distribution: Refers to a distribution in which the frequencies fall off in geometric progression as the variate values increase. The expression usually gets confined to discontinuous distributions and does not, for example, include the exponential distribution.

The continuous form of geometric distribution is sometimes called an exponential distribution.

The probability distribution p(x) is given as follows

$p(x) = pq^{x-1}$ where $0<p<1$, $q = 1-p$

It represents the probability of the first success in a series of bernoulli trials occurring at the xth attempt. Each trial has probability p of success, probability q of $1-p$ of failure.

For this distribution

$$\text{Mean} = \frac{1}{p}$$

$$\text{Variance} = \frac{q}{p^2}$$

The distribution should not be confused with that which appears normal when plotted on a horizontal logarithmic axis, of which the mode and median are the geometric mean.

Geometric mean: Refers to the average value of a set of n numbers, which is expressed os the nth root of their product. Thus, the geometric mean of n numbers $x_1, x_2, ... x_n$ is as follows:

$$\sqrt[n]{x_1 \, x_2 .... x_n}$$

Geometric moving average: Refers to an expression to be avoided, as it has been not based on geometric means.

Geometric progression: A series (or sequence) of numbers ranked in rising or falling order such that the ratios between successive pairs of terms are constant.

Example: The series 1, 3, 9, 27, 81........is a geometric progression because the ratio between each, successive pair (1, 3), (3, 9), (9, 27) is constant (= 3). The median of the series is the geometric mean (or geometric average). The sum of a geometric progression is

$$a \, \frac{r^n - 1}{r - 1}$$

where a is the first term of the series, r is the ratio between each successive pair of terms and n is the number of terms. Thus, in the above example, $a = 1$, $r = 3$ and n=5, so that the sum of the terms is

$$1 \times \frac{3^5 - 1}{3 - 1} = \frac{242}{2}$$

$$= 121$$

Among other things; the concept of geometric progression is important in compound interest calculations for the computation of sinking-fund values, where the increment of a rate of interest (r) may be expressed as $(1+i)$ being the rate of interest, and therefore the sum of a sinking fund of n terms (i.e., n-1 years) may be calculated using the formula

$$\frac{(1+i)^n - 1}{i}$$

Geometric range: This expression is found very infrequently in two senses; (i) as the ratio of the extremes of a sample, (ii) as the geometric mean of the extremes of a sample. Neither usage is having much to recommend it.

Given period: In the construction of index numbers it becomes necessary to relate data at two points of time, a base period and some other period which may be earlier or later in time, the 'given' period.

Gompertz curve (growth curve): Refers to a trend curve which is representing either growth or decline where the rate of change either increases or declines at a constant percentage or rate. The equation of the curve has been

$$Y = ae^{-bt}$$

where Y denotes the dependent variable, a denotes a constant and bt denotes the constantly increasing or declining rate of growth.

Goodman-Kruskal Tau: Refers to a measure of association for cross classification in contingency tables. It has been asymmetric between the two qualities concerned, being based on a probability interpretation involving the nation of predicting one given the other.

Goodness of fit: A measure of the agreement between observed and expected values. The expected values may be determined by a regression model (in the case of regression analysis) or some other proposed or theoretical model. The extent of goodness of fit is measured by the extent to which expected values plotted on a given line or curve do not differ from observed values. Generally the lack of fit is measured in term of some function of the sum of squared differences between observed statistics and expected values. The chi-squared test can be used to test goodness of fit in many cases.

The term is used especially in relation to the fitting of theoretical distributions to observation and the fitting of regression lines.

Goodness of fit may be measured, for grouped data, by the statistic χ^2, where

$$\chi^9 = \sum_{all\ groups} \frac{(O-E)^2}{E}$$

for the correct number of degrees of freedom, where 0 is the observed number in a group, and E the expected number.

Grade: Galton used the word to mean the relative cumulative frequency of the variate values or the proportion of total frequency having values less than or equal to the variate value. In the table under cumulative frequency, for example, 70 of the 250 office workers have salaries of Rs. 1999 or under. The grade of the value Rs. 1999 is therefore 7/25ths, or 0.28.

If the categories used in the cumulative frequency distribution are discrete, the calculation of grade is similar except that only half the frequency value of the highest variate class is included. Thus if total frequency is 1000 and variate ranges from 1 to 10, and if the cumulative frequency of variate values from 1 to 6 is 575 and the frequency of variate value 7 is 80, the calculation of grade for variate value 7 would be

$$\frac{575 + \frac{1}{2}(80)}{1000} = 0.615$$

Grand average: The average of a universe as distinct from that of a sample of the universe.

Grand mean: A mean of a number of mean values. The term is used, *inter alia*, in analysis of variance and in mean range method of computing standard deviation. Thus, in the latter usage, a group of items is divided into subsets of a specified size (e.g., $n = 6$). Then the means and ranges of each of the sets are calculated. The mean of the ranges is determined and multiplied by a factor whose value depends on the number of items used in each subset. The result is an approximation of standard deviation. The parameter mean may be, estimated by computing the mean of each of the means of the subsets. This mean of means is the grand mean.

Note that when the means are calculated from subsets of different sizes, they should be weighted to obtain the grand mean. Thus if the mean of a group of five observations is 8, and the mean of another group, consisting of fifteen observations is 12, the grand mean is

$$\frac{(5\times8)+(15\times12)}{20} = \frac{220}{20}$$

$$= 11$$

or, more simply

$$\frac{(1\times8)+(3\times12)}{4} = \frac{44}{4}$$

$$= 11$$

Graph: The presentation of statistical data in geometric form. This use of visual image facilitates the comprehension of the pattern of distribution of a variable (a characteristic of the population studied) or the relationship of two or more variables. Differences in quantities are represented by the area, shape, colour, or other differentiating characteristics of the design.

Graphical estimator: A constant which was chosen by trial and error from a geometrical representation generally for securing a linear plot of obser-vations on a specially ruled paper.

Grid: Two sets of parallel equally spaced lines drawn such that the sets of lines intersect each other vertically. A grid may consist of horizontal and vertical lines or of diagonally drawn intersecting lines. Grids may be used in area sampling to select particular subjects of an area for specific examination.

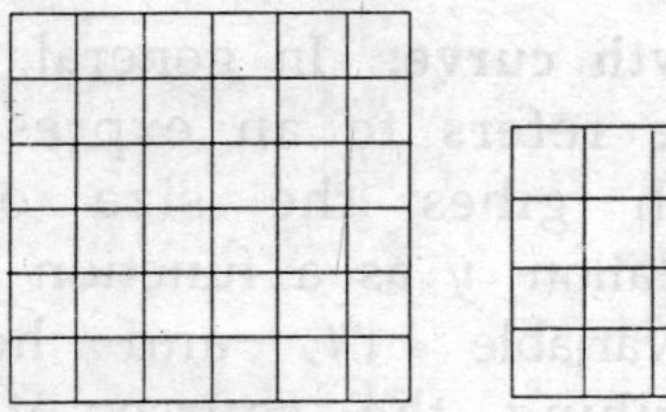

Examples of grids

Grid sampling: Refers to a form of cluster sampling, the clusters being individual areas of a grid and hence having groups of basic cells which are arranged in some standard geometrical pattern. The term 'configurational sampling' has been also used in the same sense.

Gross errors: The total errors involved in the measurement or observation process. The examination of gross errors is particularly important in methodological studies where we are interested in the process of measurement, itself. Whereas estimates of population means or totals are affected only by net errors—that part of the total error which does not cancel our over all the observations—measures of association are affected by gross errors. In addition, any attempt to eliminate errors must be based on a knowledge of the gross errors involved and the circumstances under which they arise.

Grossing up: A process used for estimating the total for a universe from a sample whose proportion of the universe is known precisely or approximately. The multiplier used for the purpose is known as the *growing-up factor*, which, for a 5% sample, would be 20.

Group: Refers to a set of elements, individuals or observations all of which possess one, or more, characteristics in common. The word is also found occasionally in statistics in its mathematical sense.

Group averages: A method of simple-regression analysis in Which the sets of observations are arrange in two or more groups of equal size in accordance with the magnitude of the independent variable averaged, and either plotted in a graph usually with the individual items, or analysed by finite differences. The term group averages is used in reference to simple regression, the term cross-classification which is also a method of group averages in multiple regression. Both are methods of regression analysis in their own right. But their chief value lies in empirical model building to provide it basic equation for use with least squares.

Group comparison: Refers to a comparison between groups of which is usually done on the basis of a representative value (such as a mean) from each.

Group factor: See common factor.

Grouping lattice: A lattice or mesh of equal variate intervals which, superimposed on a variate scale, defines the intervals in which the frequencies have been grouped.

Group internal - control matrix: A matric or table designed to facilitate the control of the inter-divisional transactions of a company or the intercompany transactions of a group. Basically, it consists of an inter-output matrix internalto the company or group.

Grouped poission distribution: In the pure birth stochastic process if it sis assumed that the time between two events is having the gamma distribution then the population size at a given time is having the grouped Poission distribution with parameters k and λ with probability function

$$\Pr(x) \sum_{j=1}^{k} e^{-\lambda} \frac{\lambda^{xk+j-1}}{(xk+j-1)!}$$

where $x = 0, 1, 2, ..., k$

has been a positive integar and λ has been positive.

Growth curve: In general, this curve refers to an expression which gibes the size of a population y as a function of a time-variable t, and hence describing the course of its growth. The expression may also be used for denoting the growth of an individual.

H

Half normal distribution: Refers to a special case of the folded Normal distribution. If the distribution of $z = |x|$ is 'folded' in half and has the density function.

$$f(x) = \frac{1}{\sigma} \sqrt{\frac{2}{\pi}}\, e\text{-}i^2/2\sigma^2.$$

Half normal plots: Refers to a graphical method for interpreting the contrasts in a two level factorial experiment. By ranking these contrasts they could be shown as a cumulative distribution function on a modified form of arithmetic probability paper. The modifi-cation consists of suppressing the lover half of the probability scale and replacing each value on the upper half by P' =2P—100. The ordinary plotting convention of $P^1 = (i—½)/n$ has been then used.

Half-normal probability paper: Normal probability paper with negative abscissas omitted.

Half-replicate design: Refers to an experimental design based upon the principle of fractional replication, which uses only one half of the complete number of treatment combination in a basic design.

Half-width: This expression has been sometimes used in relation to central co ıfidence intervals for denoting the upper or lower half of an interval. In fixed interval prediction, such as may be used in a control chart, the half-width refers to the distance on the scale of the variate between, say, the process average and the upper or lower control limit.

An objectionable use of the term sometimes takes place in elementary texts. If a frequency function has been unimodal with a frequency f_0 at the mode, and if there have been two points x_1 and x_2 where

$$f(x_2) = f(x_2) = ½ f(x_0)$$

the distance ½ $(x_2—x_1)$ has been sometimes called the half width.

Haphazard sampling: Selecting been sometimes called the halfwidthout taking particular a care about the representativeness of the elements selected. Samples of volunteer subjects are haphazard whole population on the basis of such sample. The term should not be confused with random sampling in which every element in the population is having a known non-zero probability of selection.

Hard currency: A currency which has a continuing high level of demand, relative to supply in the market for foreign exchange. Currencies of stable industrialized nations can usually be included, whereas those of *underdeveloped countries* and the Eastern Bloc are not.

Hardy summation method: Refers to a method of determining the moments of a frequency function defined at equidistant points, or grouped in equal intervals, by repeated summation of the frequencies. It was given by Sir George Hardy in 1903. The summations provide factorial moments from which ordinary moments may easily get derived.

Harmonic analysis: Refers to the analysis of a series of value, into constituent periodic terms.

Harmonic mean: A special kind of mean which is determined by calculating the mean of the reciprocals of the values concerned and then calculating the reciprocal of this quantity.

The harmonic mean *m* of the *n* numbers x_1, x_2, ..x_n, is given by:

$$\frac{1}{m} = \frac{1}{n}\left\{\frac{1}{x_1} + \frac{1}{x_2} + \ldots + \frac{1}{x_n}\right\}$$

Thus, the harmonic mean of 2, 4 and 5 may be calculated:

$$\frac{1}{m} = \frac{1}{3}\left(\frac{1}{2} + \frac{1}{4} + \frac{1}{5}\right) \text{ giving}$$

$$m = \frac{60}{19} = 3\frac{3}{19}$$

The harmonic mean may be thus also given by:

$$\frac{n}{\sum_{i=1}^{n} \frac{1}{x_1}}$$

Harmonic process: A stationary random process which satisfies the relationship:

$$x_1 + b_1 x_{t-1} = b_2 x_{1-2} + \ldots + b_h x_{1-h} = 0$$

has been a harmonic process. It is deterministic, the general solution being a sum of harmonic terms which may get damped or explosive.

Harris walk: Refers to a random walk on the set of non- negative integers, for which the matrix of transition probabilities is consisting of zeros except for the elements.

$$P_{0,1} = 1$$

$$P_{j',\ j+1} = a_j \quad 0 < a_j < 1,\ 1 < j < \infty$$

$$P_{j',\ j-1} = 1 - a_j$$

Hazard: In general, refers to a word which implies the existence of chance of risk (from Arabic 'al zhar', meaning a die). Specialised usage takes place principally in connection with lief analysis of physical systems or components, The hazard has been the probability that an article, functioning at time t_0, will fail in the interval(t, t+δt) and this is given by $h(x)= f(x)/[1-F(x)]$. If this instantaneous (death or failure) rate has been calculated at a number of survival times, the resulting hazard rates may be considered as a hazard function.

Helmert distribution Refers to the distribution of the sample standard deviation or, equivalently, of the sample variance in samples from a normal population.

Hermite distribution: The probability generation function of a generalised Poisson distribution may be put as g (z)=

$$\exp\{a_1(z-1) + a_2(z^2-1) + ...\}$$

where $\sum_{i=3}^{\infty} a_1 = \lambda$, the Posson parameter. This generalised distribution tends to become the Poisson' distribution having parameter a_1 if a_2, a_3... as become negligible compared to a_1 in some limiting process. If a_2 could not be neglected as compared to a_1 then limiting distribution would be then Hermite distribution with parameters a_1, and a_2.

Heteroscedasticity: Scedasticity is a term, which is not much used. It denotes dispersion, particularly in terms of variance. The term heterosce-dasticity finds use in regression analysis to describe a situation where the variance of the error or disturbance term differs for different values of the independent variable. In general, the distribution of a variable has been said to be hetero-scedastic if its variance differs for different fixed values of another variable. If the variance has been the same for all fixed values of the other variable, the distribution has been said to be homoscedastic.

Heterotypic: Refers to a term which is used in relation to -Pearson distributions. For certain values of the moment ratios $\beta_1$and β_2 the differential equation defining the family of distributions is having infinite moments of orders 8 or more. In this region the standard error of the sample estimate of β_2 would be infinite and hence the fitting of the distributions by the method (if moments would get inappropriate. Distributions of the Pearson family with such values of β_1 and β_2 had been termed as hetrotypic; but, as is now realised, certain of the distributions may get valuable for many purposes.

Hierachy: If, in a matrix of intercorrelations of a set of variates the rows and columns could be so arranged to provide the highest correlations in the upper left-hand corner and the lowest correlations in the lower right-hand corner and when this is Performed there has been a constant proportional relationship between adjacent columns, except for diagonal terms, the table has been called a hierarchy and the intercorrelations are said to be an hierarchical order. Thus for two rows denoted by a and b an two columns and *d*, the correlations are obeying the so-called tetrad relations $r_{ac}\ r_{bd} = r_{ad}\ r_{bc}$. Under certain conditions the fact that the correlation matrix has been hierarchical has been a necessary and sufficient condition that the variation can be accounted for by a single- factor common to the variates.

The quantities $r_{ac}\ r_{bd} - r_{ad} r_{bc}$ are termed as tetrade differences.

High contact: In relation to a frequency function $f(x)$ of contact of the function having the variate axis or at infinity if the range has been infinite, has been said to be high if $x^{rf}(x)$ or its limit vanishes at the terminals for some high value of *r*. What constitutes a high value for this purpose has been some what arbitrary. This Property of high contact has been one of the conditions for the application of the corrections for grouping known as Sheppards corrections.

High powered money: In the traditional theory of the credit multiplier, the reserve assets which form the base on which the banking system creates bank deposits, and which constrain the lending activities of the banks which lead to deposit creation, are collectively termed 'high-powered money' since, according to the theory, a change m the quantity of these assets will produce a multiplied change in the bank deposit component of' the money stock. The term also refers to the fact that the clearest constituents of high- powered money are the monetary liabilities of the public sector, i.e. circulating currency and bankers' balances at the central bank. However, short-term assets, e.g. certain bills which can readily be converted into public sector monetary liabilities at the central bank should arguably be included if they are treated as reserve assets by the commercial banks, it' bank lending is influenced by the banks' holding of them, and if their supply is an object of central bank control.

Histogram: In statistics, a form of graphic presentation of a frequency distribution, or distribution of cases, in which a vertical bar is constructed for each class interval, or sub d of the variable being presented. The class intervals are arranged along the horizontal axis and *the* width of each bar corresponds to the size of the class interval. The frequencies are arranged along the vertical axis and determine the height of the bars. That is, the height of each bar is determined by the number of cases in each class interval. A histogram is similar to a bar graph except that there are no spaces between the bars.

Historigram: The graph of a time series with the series measured on the vertical scale. It is common practice to connect adjacent dots by straight lines and not attempt to fit a smoothed curve through the dots themselves, through a general curve can be distance between scale points on the time scale represents the lapse of time rather than periods, so that lines drawn from the calibration marks of the time scale should represent either periods or points of time for rate- of-flow figures and point-of time figures, respectively ; the plotted dots thus fall on the lines and not between' them. Owing to possible confusion with the word 'histogram the term is not to be recommended.

Holding company: A company that has control over a number of other companies, through ownership of a sufficient proportion of the latters' common stock. A sufficient proportion is 51 per cent or over, however, where there is a dispersed and inert group of stockholders *minority control* can be exercised with a proportion of the common stock which is below 51 per cent. This enables holding companies to acquire subsidiaries at a cost which is less than could have been the case if it had purchased their physical assets. This pyramiding of control, as it is called, is subject to limitations imposed by company legislation.

A holding company tends to concern itself' with control over the financial, managerial or marketing functions of its subsidiaries, and thus performs more of a monitoring role. The subsidiaries are then given an element of independence, which facilitates the retention of the individual subsidiary's image and control over its production and investment plans.

Homogeneous/homogeneity: The term has been used in statistics in relation to samples from different populations which may or may not be identical. If the populations have been identical, they are said to be homogeneous, and thus the samples drawn from these populations are also said to be homogenous, It has been more common to restrict the examination to particular characteristics of the populations, *e.g.,* the means or variances. There a number of populations with identical means are homogeneous in their means.

Homogeneous process: A stochastic process is considered to be homogeneous in space if the transition probability between any two state values at two given times is depending only on the difference those state values.

The process has been homogeneous in time if the transition probability between two given state values at any two times depends only on the difference between those times.

Homograde: A term used to denote a qualitative variate i.e., an attribute. The word has been not in common current use among English writers.

Homoscedasticity: Scedasticity is a term, not widely used, which denotes dispersion, particularly in terms of variance. The term homoscedasticity is used in regression analysis of situation where the variance of the error term or d describe a term is the same for all values of the independent variable. In general, the distribution of a variable is said to be homoscedastic if its variance is the same foe all fixed values of another variable. If the variance differs for different values of the other variables, the distribution is said to be heteroscesatic.

Hunt-stein theorem: Refers to a theorem concerning critical regions which are associated with a particular significance test. For example, in the set of critical regions of size a which have been also invariant, there may be one, say, W_0, which has been uniformly most powerful and most stringent.

Hybrid series: A statistical series consisting of 'a mixture of time series figures and cross-section figures, sometimes called more precisely a temporal crosssection series. It is adopted for, analytical purposes where (1) enough data of one kind or the other are not available (2) cross-section data are limited in number of observation and have to be supplemented by time data; (3) in a regression preferred but are available for some variables only, and time series data have to be used for other variables. In the latter case, two separate analyses are carried out, and the two resulting regression equations merged into one.

Hyperbolic secant distribution: Refers to a frequency distribution which gets related to the logistic distribution, with probability density function

$$Pr_y\,(y) = \pi^{-1}\ \text{sech}\ y$$

and cumulative distribution function

$$F_y(y) = \tfrac{1}{2}+\pi^{-1}\ \tanh^{-1}\ (\sinh\ y)$$

Hyperexponential distribution: A term which was proposed by Morse to a mixture of exponential distributions in the general form :

$$\gamma e^{-2\gamma bt} + (1-\gamma)\, e^{-2(1-\gamma)bt.}$$

Hypergeometric waiting time distribution: Refers to an alternative name for the inverse hypergeometric distribution.

Hyper-poisson distribution. This discrete distribution is having a complicated probability function in two parameters, λ and θ, If $\lambda = 1$ this distribution becomes the Poisson distribution with parameter θ. If $\lambda < 1$ it has been sometimes referred to as a sub-Poisson distribution and if $\lambda > 1$ as a super-Poisson distribution.
where

$$c_p = \left[\int_0^{\pi} (\sin^{p-1} t) \exp(k \cos t)\, dt\right]^{-1}$$

Hypothesis: In scientific investigation, a tentative statement asserting a relationship between certain facts. The statement is intended to be tested empirically and either verified or rejected. A scientific hypothesis is derived from a the theoretical system and the results of past research. The clear statement of a theoretically significant hypothesis related to previous research is crucial to a well-designed scientific study.

Hypothesis, statistical A hypothesis which is concerning the parameters or form of the probability distribution for designated population or populations, or more generally, of a probabilistic mechanism which has been supposed to generate the observation.

Hypothetical population: Refers to a statistical population which is having no real existence but is imagined to be generated by repetitions of events of a certain type e.g., the binomial distribution as generated by the throws of a die, or crop-yields on a set of plots imagined as all the possible ways in which a set of yields could occur under the conditions of an experiment.

I

Ideal index number: In 1922 Irving Fisher put forward certain criteria which have to be obeyed by 'good' index numbers. Of the large collection of formulae investigated only a few have been found to obey his text.

It may be but as follows:

$$\left\{\frac{\sum p_n q_o}{\sum p_o q_o} \times \frac{\sum p_n q_o}{\sum p_o q_n}\right\}^{1/2}$$

where p_o, q_o denotes prices and quantities in the base period and p_n, q_n those of the period for which the index is being calculated. The ideal index has been the geometric mean of the Laspeyres' and Paasche index numbers.

Identical errors: A regression model of type $y = f(x_1 ..., x_p) + e$ where e posseses an unique frequency distribution, is considered to have identical errors. 'Identity' refers not to individual errors but to their frequency distribution.

Illusory association: An association between attributes may be sometimes said to be illusory if it may be statistically significant without necessarily involving any direct causal connection between them.

Imbedded process: A stochastic process in continuous time has been observed only at the time points where a change of state takes place. These points of discontinuity can be considered to be forming a new discrete time variable.

Imperfect competition: A degree of competition which affords the seller an appreciable amount of control over his selling Prices.

Implied elasticity of demand: A null hypothesis, to the effect that the current price of a brand of a firm's products is rational based on the postulation that the implied price elasticity of demand is the true one. The formula of the rational price or optimum price, as it is often called, as

$$p_m = a \frac{e}{e-1}$$

where p_m is the rational price, a the marginal cost, and e the elasticity of demand.

Implicit strata: Refers to the *n* groups, zones or subclasses into which, in systematic sampling, the population has been divided by the sampling interval $k = N/n$, where N denotes the population size and n the sample size.

Importance sampling: This method of sampling was used in Write Carlo analysis and was used to concentrate the distribution of sample points in those parts of the variate interval which have been of greatest interest. It has been a form of stratified sampling having variable sampling fraction.

Impossible: Incapable of occurring of happening. If an event is impossible, the probability of its occurrence is zero.

Improper distribution: Refers to a probability distribution having a distribution function F(x) which $F(-\infty)>0$ or $F(+\infty)<1$. Such distribution are having adherent probabilities at $-\infty$ or $+\infty$ respectively.

Incidence rate: The number of spells beginning during the observation period rated to the average number of persons under observation.

Income consumption curve: The tangency of the consumer's indifference curve and his budget line defines the consumer's equilibrium. If income is varied such that the budget line moves outwards in a parallel fashion, new equilibrium will be established. The locus of all such equilibrium is the income consumption curve. If the curve slopes upwards in commodity *space it* indicates that both goods are *normal goods.* If it slopes backwards or downwards one of the goods is an *inferior good.* The income consumption curve is widely referred to as an *Engel curve,* although some writers distinguish the two.

Income effect: A change in the price of a good will reduce or increase the consumer's real income. In response to this change in real income we can expect the consumer to buy less (more) of all goods including the one which has changed in price. This is the income effect of a price change.

Income elasticity of demand: A measure of the responsiveness of the quantity demanded of any *good* to a change in the level of income of the persons demanding the good. To avoid the measure being sensitive to the units used, the measure is expressed as:

Income—expenditure model: A simple Keynesian one sector model which shows the determination of the *equilibrium level of national income.* Under the assumptions of a fixed *capital stock, labour force* and technology and rigid wages and prices and, therefore, expectations, it details the static equilibrium in which the supply of *real national output* equals the quantity of national output people wish to buy. In the model the volume of *aggregate expenditure* determines the volume of output and the associated level of employment: the model concentrates exclusively on the demand side of the macro-economy. There is no reason why the equilibrium that is eventually established will be associated with full employment, on the contrary this simple representation of the Keynesian system seeks to demonstrate the possibility of an *unemployment equilibrium.* The equilibrium level of national income is established where *withdrawals* equal *injections* and this may occur at less than full employment. Under these circumstances the Keynesian prescription is for a rise *in government expenditure* which when magnified by the actions of the *multiplier* will result in a level of national income and output just sufficient to ensure full-employment. Graphically equilibrium is shown as the point at which the *aggregate expenditure* function cuts the 45′ line as below.

This simple approach can however, be criticized on a number of counts. First, it concentrates on the demand side of the economy assuming that the appropriate supply will be forthcoming at any level of expenditure. Second, money plays no role in the model; prices are determined exogenously. The model in fact concentrates exclusively on the goods market and it is the ISLM *diagram which* integrates this with the money market.

Incomplete beta function: This function may be defined as follows;

$$B_t\ (s,\ r = \int_0^t y^{s-1}\ (1\text{-}y)r^{-1}dy.$$

$$s^t r>0;\ 0 \leq t \leq 1.$$

Incomplete coverage: A survey or census in which a considerable number of the elements in the population are not included may be called incomplete.

Incomplete gamma function: A function which is defined as follows:

$$\lambda_t\ (\lambda) = \int_0^t e^{-x} x^{\lambda-1} dx\ \ \lambda>0\ ;\ 0 \leq t \leq \infty$$

It has been the distribution function of the gamma distribution multiplied by λ(λ).

Incomplete moment: The ordinary moment of a distribution about an arbitrary origin *a* may be given by

$$\mu_r \int_{-\infty}^{\infty} (x\text{-}a)^r \quad dF(x).$$

Increment 1: Refers to a small change in the value of a variable, 2. Refers to change the value of a variable system actually by small amounts.

Incremental analysis: Refers to the process of determining an optimum value by building up a statistical schedule.

Independence: This term in the calculs of probabilities, may be generally defined by reference to the principle of compound probabilities. Two events are considered to be independent if the probability of one has been the same whether the other has been given or not, i.e., $Pr(A) = Pr(A \mid B)$ and $Pr(B) = Pr(B \mid A)$.

From this it is evident that the probability of the compound events are independent. As a matter of axiomatization it is desirable use relations of the type $Pr(AB) = Pr(A)\ Pr(B)$ as definitions to avoid difficulties arising when $Pr(A)$ or $Pr(B)$ is zero.

In statistics two variates x_1 and x_2 have been independent if their distribution functions get related by

$$F(x_1, x_2) = F(x_1, \infty)\ F(\infty, x_2)$$

or equivalently if their frequency functions, should they exist, are related by $f(x_1, x_2) = f_1(x_1)\ f(x_2)._2$ Generally n variates $x_1, x_2 \ldots x_n$ have been independent if

$$F(x_1, x_2, \ldots x_n) = F(x_1 \infty, \infty, \infty, \ldots \infty)$$

$$F(\infty, x_2, \ldots, \infty) \ldots F(\infty, \infty, \infty, \ldots\ x_n)$$

It has not been enough that they should be independent pair and pair. The word has been also applied in the ordinary mathematical sense to describe the independence of two or more variables.

Independence frequency: In a contingency table, this term refers to the frequency which would be found in a particular cell if the attributes defining it had been independent ; e.g., if the rows of the table are having total frequencies $A_1, A_2 \ldots, A_r$ and the s columns are having frequencies $B_1, B_2, \ldots B_s$ and

$$\Sigma A = \Sigma B = N,$$

the independence frequency in the ith row and jth column would be $A_i B_j / N$.

Independent events: Refer to the events which have no influence on each other. If a die and a coin have been thrown together, the events 'The die shows 6' and 'The coin comes heads' have been independent.

This is expressed mathematically by saying that event A has been independent of event B if

$$p\ (A \mid B) = p\ (A \mid B') = p\ (A)$$

In the same way, event B is independent of event A if

$$p\ (B \mid A) = p\ (B \mid A') = p(B)$$

In fact, if A is independent of B, then B must be independent of A and

$$p(A \cap B) = p(A)\ p(B)$$

Independent random variables: Refers to the random variables, e.g., x and y, such that a knowledge of the value of x does not inference the probability distribution of y, and *vice versa*. Thus there is no relationship between the values of independent random variables, and their covariance is zero.

Independent, trials: The successive trials of an event have been considered to be independent if the probability of outcome of any trial has been independent of the outcome of the others. The expression has been usually confined to cases where the probability has been the same for all trials.

Independent variable: This term is used in regression analysis to mean one or a number of predictor or explanatory variables (or variates) from which a dependent variable is said to have resulted. It is a convention to use the letter Y to indicate the dependent variable and $X_1, X_2, X_3, \ldots X_n$ for each of the independent variables. Thus a relationship between independent variables may be defined

$$Y = \beta_0 + \beta_1 X_1 + \beta_2 X_2 + \beta_3 X_3 + \ldots + \beta_n X_n + \varepsilon$$

Index: Any measurable or observable phenomenon (or phenomena) that is used to indicate the presence of another phenomenon that cannot be measured directly or conveniently. Thus one may attempt to determine the social status of individuals in a social system by determining their occupations. Occupation, then, is used as an index of social status.

Index numbers: A statistical series denoting relative values, usually where there is no absolute measure, and almost invariably standardized to one item of the series given as 100. The time or place given as equalling 100 is called the base.

In Laspeyre's Index the weighting has been taken from the base year, in Paassche's Index from the current year. This implies that Laspeyre's Index tends to emphasize items which are out of date and to ignore changes in taste or life style over the years. Paasche's index, on the other hand, runs into difficulties with new items which are significant in the current year but have no base value because they did not exist or were not used before. Imagine, for example, a cost of living index, set up with base year 1800 to compare the cost of living then with now. If Laspeyre's index has been used, items like candles and horse feed will be quite heavily weighted, whereas electricity and petrol would not feature at all. If Paasche's index has been used, the reverse has been the case.

A compromise between the two has been the Typical Year Index, where the weightings have been taken for some typical time between the base year and the present.

In Fisher's Ideal Index, the value has been taken as the geometric mean of Laspeyre's Index and Paasche's Index. The Marshall Edgeworth Bowley Index takes the weighting for any item to be the average of that for the base year and that for the current year.

Formulae for the various indexes are as follows, where p_o=base year price, p_n= current year price, q_o=base year weighting, q_n= current year weighting, and q_b= typical year weighting:

$$\text{Laspeyre's Index} = \frac{\Sigma p_n q_o}{\Sigma p_o q_o}$$

$$\text{Paasche's Index} = \frac{\Sigma p_n q_n}{\Sigma p_o q_n}$$

$$\text{Typical Year Index} = \frac{\Sigma p_n q_t}{\Sigma p_o q_t}$$

Fisher's Ideal Index=

$$\sqrt{\left\{\frac{\Sigma p_n q_o}{\Sigma p_o q_o}\right\}\left\{\frac{\Sigma p_n q_n}{\Sigma p_o q_n}\right\}}$$

Marshall Edgeworth

$$\text{Bowley Index} = \frac{\Sigma p_n\ (q_o+q_n)}{\Sigma p_o\ (q_o+q_n)}$$

Indirect effect: Refers to that part of the effect of one variable of another in a path model which operates through its effect of an intermediate variable in the model.

Indirect sampling: Refers to sampling from documents, or some record of the characteristics of a population, rather than the recording of information obtained at first hand from units of the population themselves.

Individual response error: The individual the difference between the individual true value and the information recorded for the individual.

Inefficient statistic: Refers to a statistic having less than the maximum Possible, efficiency in the sense of having the minimum sampling variance.

Inferential statistics: Refers to that branch of statistics which involves drawing conclusions from data by hypothesis testing.

Infinite divisibility: A distribution F has been infinitely divisible if and only if for each n it could be represented as the distribution of the convolution

$$S_n = X_{1,n} + .. X_{n,n}$$

of n independent random variables having a common distribution F_n

Infinite population: An infinite population may be defined as one which either has the infinite property through some limiting process or, in a sampling context can be made to possess that property by some strategy of sampling.

Inflation accounting: Refers to techniques for dealing with the impact of inflation on accounts and accounting procedures. Traditionally, historic cost methods have been employed whereby accounts were built up from actual records of transactions. In periods of rapid inflation the replacement cost of fixed assets would greatly exceed historic costs. Consequently in real terms depreciation will be understated when historic cost accounting is employed and profits overstated.

The system of capital allowances may then not permit the recovery of capital costs in real terms. Similarly, rapid inflation can give an artificial boost to profits from stock appreciation. If such profits are taxes a company which has to replace its stock at the higher price level is penalized. Several different approaches to a solution to the problem have been suggested.

The current purchasing power method retains historic cost notions but expresses accounts in terms of purchasing units' employing a retail price index. The Sandilands Committee which examined the whole subject in the UK recommended the use of value accounting whereby assets were to be measured by reference to their value and not cost. The use of replacement cost accounting whereby assets are revalued from historic to current costs was recommended. There has been much controversy about the detailed procedures that should be adopted. Eventually a consensus was reached with the publication of Statement of Standard Accounting Practice 16 entitled 'Current Cost Accounting' in 1980. Three adjustments to trading profits calculated on a historic basis are required to arrive at current cost operating profit.

The first is the *depreciation* adjustment which allows for the impact of price changes when determining the charge against revenue for the part of fixed assets consumed in the period. The second is the cost of sales adjustment which allows for the impact of price changes when determining the change against revenue for stock consumed in the period. The third is the monetary working capital adjustment which represents the amount of additional (or reduced) finance needed for monetary working capital as a result of changes in the input prices of goods and services employed in the business.

Inflationary gap: Aggregate expenditure exceeds the maximum attainable level output with the result that there is upward pressure on prices. In the simple income *expenditure* model this is shown by an aggregate expenditure function which cuts the 45', where E = Y, beyond the full employment level Yf . At Yf the sources of capital and labour are full employed with the result that this represents the maximum attainable level of output. As a result there exists an inflationary gap of magnitude AB with the associated upward pressure on prices.

This was initially advanced to explain the inflation of the 1940s and 1950s: the first real inflation since the publication of the general theory of employment, interest and money. However the treatment neglects the role of money either, in bringing inflation under control in exacerbating an upward spiral of' wages and prices. The model may be considered a *demand-pull* explanation of inflation.

Information: A term which is having a particular meaning in the theory of estimation.

The amount of information about a parameter ϕ from randomly drawn sample of n independent observations, where the population from which the sample is drawn has a frequency function $f(x, \phi)$ can be defined as

$$n\varepsilon \left(\frac{\delta \log f}{\delta\phi} \right)^2 \equiv n \int_{-\infty}^{\infty} \left(\frac{\log f(x.\phi)}{\delta\phi} \right)^2 f(x,\phi)\, dx$$

In general it may be said that if the variance of unbiased estimators of ϕ is large the information is small and if the variance is small the information is large.

Information theory: A mathematical theory of communication which is concerned with problems relating to the accuracy efficiency, and effectiveness of communication. It deals with questions such as the extent to which the symbols used in communication convey the intended meaning, the effect of extraneous disturbances (referred to as "noise") in introducing errors in communication, the extent to which there is "redundancy" in communication—that is portions of a message not needed to convey the intended meaning.

Input-output analysis: A technique in which the output of each industry in an economy is analysed according to the use made of it. It can be used to calculate the level of output in different industries necessary to sustain a given pattern of final demand.

Input/output process: Refers to a broad class of stochastic processes, the stale $\xi(t)$ at time t being the resultant of two random variables x, y;

$$\xi(t) = x(t) - y(t)$$

x and y being called respectively the input and output. Examples have been queueing processes, provisioning and epidemic processes.

Inspection lot: A lot presented for inspection, which could be carried out on each member of the lot or on a sample of members only.

Instalment credit: The general term for finance lent on conditions which provide for repayment of the principal, together with the *interest, by* regular instalments. The greater part of' instalment credit is advanced to finance the purchase of consumer goods under *hire purchase* contracts, where the finance is linked to a particular transaction with the lender and purchaser having certain legal rights in the object purchased.

Such credit may be financed by the retailer concerned, although in the case of hire purchase the contract may be between the buyer and a *finance company.* Other forms of instalment credit include *personal loans* by banks, and credit extended through *credit card* and budget account facilities. As normally used the term does not embrace mortgage finance for house purchase. It should be noted that, increasingly, producer goods are purchased under various instalment credit arrangements.

Instantaneous death rate: Refers to a version of the age specific death rate where the next unit time period has been reckoned to be short.

Instantaneous state: Refers to a state of a Markov chain in continuous time for which the expected sojourn time has been zero.

Integer programming: Refers to a form of linear programming where the solutions get derived in an integer form, e.g. In the scheduling of ships or men.

Integrated data: Refers to a class of statistical data in which the values for short unit intervals could be added together to provide a series of values relating to longer intervals.

Intensity: In the harmonic analysis of time series, this term refers to a measure which gives an estimate of the amplitudes of the constituent harmonics.

Intensive sampling: Like extensive sampling this expression may imply two different things: either (a) sampling in a particular area with a dense scatter of sampling points, or (b) sampling where in information on a restricted range of topics has been sought by probing on them very deeply with an intricate schedule of questions.

Interaction: When the effect of a change in one variable differs for different levels of another variable there is said to be an interaction between them,

Intercorrelation: A term which is used to denote the correlation distinct from of a number of variates among themselves, as the correlations between them and an 'outside' or dependent variate.

Interdecline range: The range of observations of a variate between the first and ninth deciles.

Intermediate products: Goods at an intermediate state of production-neither raw material as purchased by a manufacturing firm, nor the finished article ready for sale to customers; an important concept in the censuses of production.

Intergraduated values: Values which have been subject either to cograduation or to contragraduation.

Internal Rate of Return (IRR) That rate of discount which converts the total yield gross of depreciation—of a capital project, to a present value equal to the capital outlay on the project.

Interpenetrating samples (subsamples): With interpenetrating sampling, instead of selecting a single sample from the population we select a number of subsamples using the same sample design for each. Each subsample must itself be a self-contained sample of the population. Interpenetration may be used with any sample design and at any stage of the selection.

Interpolation: The process of calculating an unknown value of a variable that is between two or more known values in the series. When the known values are plotted on a graph, the process of interpolation assumes that the unknown value lies on the line of the graph.

Interpretive statistics: Refers to that branch of statistics which is concerned with drawing conclusions from data.

Interquartile range: Refers to the variate distance between the upper and lower quartiles. This range is having one half of the total frequency and provides a simple measure of dispersion which finds use in descriptive statistics.

Interrupted distributions: A name assigned to distributions which are arising in circumstances in which some values of a random variable cannot be observed, e.g., radioactive particles arriving in the 'dead' time during which the counting instrument, having just recorded the arrival of a particle, has been unable to record further arriving particles. The distribution of particles recorded therefore is different from the distribution of actual arrivals.

Intersection: Refers to the elements that have been common to two or more sets (Fig). The intersection of sets A and B has been written A∩B. For example, if A {8, *t*, *a*, *r*} and B = {*p*, *l*, *a*, *n*, *e*, *t*} then A∩B = {*a*, *t*}

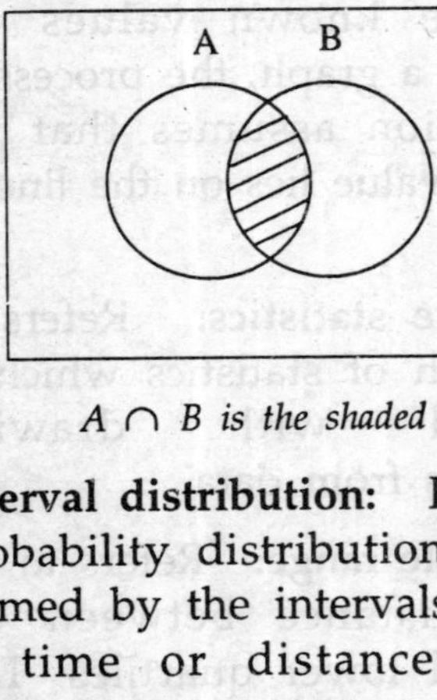

A ∩ B is the shaded area.

Interval distribution: Refers to a probability distribution which is formed by the intervals, or gaps, in time or distance between events or occurrences.

Interval estimation: Refers to the estimation of a population parameter by specifying a range of values bounded by an upper and a lower limit, within which the true value has been asserted to lie, as distinct from point estimation which assigns a single value to the true value of the parameter. The unknown value of the population parameter has been presumed to lie within the specified interval either on a stated proportion of occasions, under conditions of repeated sampling, or in some other probabilistics sense.

Interview: As structured social interaction between two people—an interviewer whose task it is to obtain information, and a respondent whose function it is to provide this information.

Interviewer effect: A term sometimes used to denote the totality of errors in survey responses due to the interviewer. It includes therefore interviewer bias and interviewer variance.

Intraclass: Within a class or group. We distinguish between intraclass and interclass. In the analysis of variance, for example, the total sum of squares can be partitioned into the interclass (as between groups) sum of squares and the interclass (or within group) sum of squares.

Intraclass correlation: Refers to a measure of correlation within the members of certain natural groups or 'families'.

Intraclass correlation coefficient: A measure of the internal homogeneity of a class, category, group of family of elements.

Intracluster correlation coefficients: A synonym for intraclass correlation coefficient. The term 'intracluster' is used mainly in connection with survey sampling where the 'cluster' are the groups of elements selected in cluster sampling. The coefficient is usually positive for human populations, since people living in the same neighbourhood tend to be more similar to one another than they are to people living in other neighbourhoods. A high positive intracluster correlation coefficient implies that within-cluster variability is relatively low-the clusters are relatively homogeneous.

Intraclass variance: In variance analysis, where the data have been classified into groups, the total variation (sum of squares about the grand mean) could be expressed as the sum of two components, expressing variation among for means of groups and variation within groups.

Intrinsic accuracy: The intrinsic accuracy of a distribution with frequency function $f(x, \theta)$ may be defined as follows:

$$I = \int_{-\infty}^{\infty} \left(\frac{\partial \log f}{\partial \theta}\right)^2 f(x)\, dx.$$

Strictly, speaking it should be related to the particular parameter θ entering into the differentiation.

Ising-Stevens distribution: If there have been n_1 and n_2 objects of two kinds C_1 and C_2 arranged at random in $n = n_1+n_2$ positions along a circle then the distribution of the number of runs of either object would be as follows:

$$\Pr(x) = \binom{n_1}{x}\binom{n_2-1}{x-1}\bigg/\binom{n_1+n_2-1}{n_2}$$

$$= \binom{n_1-1}{x-1}\binom{n_2}{x_2}\bigg/\binom{n_1+n_2-1}{n_1}$$

Invariance: This term is mostly used to denote does not get changed by a particular Iran example, an orthogonal transformation of a se normal variates leaves the properties of independence and normality unaffected; they have been invariant under the transformation.

Invariance method: Refers to a principle of estimation or hypothesis testing which needs an estimator or hypo to remain invariant if the data of the problem transformation.

Inverse distribution: This expression has been senses, for example :

(i) In a sense to be avoided, to denote the sample numbers where sampling gets continued until predeter-mined number of successes is attained, e.g., the 'inverse hypergeometric has been the distribution of sample sizes n required to obtain k successes in sampling without replacement from a finite population.

(ii) an equally regrettable sense, to denote (mainly discrete) where the frequencies have quantities.

(iii) instead of expressing the distribution in terms of the variable x, to denote expansion of $F(x)$.

It has been probably safer to avoid the expression altogether.

Inverse moment: If the rth moment may be defined as $x(Xr)$. A similar relationship is existing between the rth factorial moment and its inverse.

Inverse multinomial sampling: Refers to a form of sequential analysis in which sampling gets terminated when one or other of a specified subset of classes has occurred k times.

Inverse probability: Refers to the probability approach which endeavours to reason from observed events to the probabilities of the hypotheses which could explain them, as distinct from direct probability, which reasons deductively from given probabilities to the probabilities of contingent events. A principal theorem in this connection has been Bayes' theorem. The word 'inverse' usually inverse in some logical relationship but is sometimes used in the sense of 'prier in time'. For example, in connection with the analysis of stochastic processes, it has been sometimes desirable to consider the past history of a system rather than its future development. The past probabilities of transition which have given rise to the present state have been thus sometimes called 'inverse'. This usage has been comparatively rate and not to be recommended.

Inverse sampling: Refers to a method of sampling which needs that drawings at random shall be continued until certain specified conditions dependent on the results of those drawings are fulfilled.

Inverse serial correlation: A name sometimes, used for denoting the process of attempting to determine from a set of serial correlations the series which generated them. The problem has been in fact indeterminate as different series may give the same autocorrelations.

Inversion: If we define a standard order of elements (e.g., the natural order of numbers 1, 2, 3, 4, 5 ...)as inversion of two elements means that they are in the opposite order to that given in standard order. Thus, compared to 1, 2, 3, the order 1, 3, 2 contains one inversion (3, 2 instead of 2, 3). The concept has been used in the calculation of kendall's tau.

Investment criteria: Refers to the several techniques of converting the yield as an asset to a capital sum, or the capital outlay to an annual cost, so that they can be added together or deducted one from the other; also the several different ways of expressing the net yield, more precisely called the yield criteria.

Isometric chart: Refers to a chart which tries to depict three-dimensional material on a plane. It has been a, form of axonometric chart where, the distances on the three axes have been measured on an equal scale.

Example. A common use of isometric charts has been in the economics of production, where the two dimensions comprising the 'floor' of the box may be employed to illustrate two inputs, such as man-hours and materials and the Vertical axis has been used to illustrate the volume of output of one product. The following isometric chart reveals levels of production of a printer, whose two Major inputs have been labour and paper, and whose one major output is measured in quantities of books of a standard size. The shading illustrates the position where L units of labour and P units of paper have been used to manufacture B units of books.

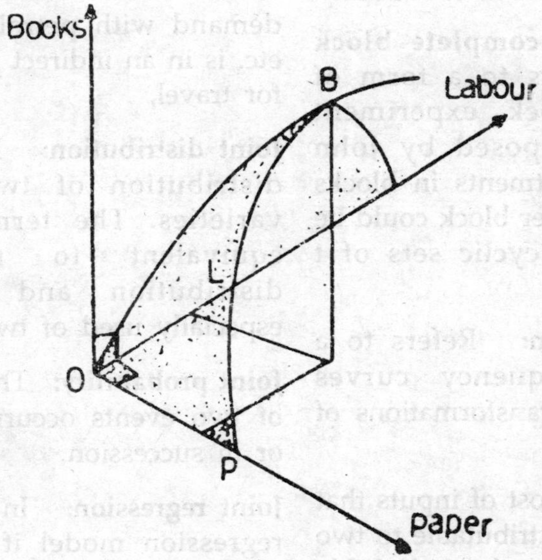

Irreduicible Markov chain: A Markov chain is considered to be irreducible if all pairs of states of the chain communicate, so that the chain is having exactly one communicating class.

Irregular variation: Refers to that variation within a time series which has been not accounted for by trend, seasonal components or cyclic components.

J

Jittered sampling: This term is sometimes used in the analysis of time series for denoting the sampling of a continuous series where the intervals between Points of observation have been the values of a, random variable.

John's cyclic incomplete block designs: Refers to a form of incomplete block experiment which was proposed by John where the t treatments in blocks of k treatments per block could be divided into k cyclic sets of t blocks each.

Johnson's system: Refers to a system of frequency curves which is upon transformations of the variable.

Joint cost: The cost of inputs that are inseparably attributable to two or more products; the indivisible part of the cost of joint products, such as coal-gas, gas-coke and sulphate of ammonia; beefsteak and sirloin; petrol and diesel oil.

Joint demand: Said of products and services that are irk demand one with another, Petrol for motor vehicles, lubricating oils and tyres are said to be in joint demand ; so, too, are bread and butter, The joint demand for petrol, oil and tyres for cars is derived from the demand for travel, which is described as being in joint demand with petrol, etc. Petrol, etc. is in an indirect joint demand for travel,

Joint distribution: Refers to the distribution of two or more varieties. The term has been equivalent to multivariate distribution and has been especially used of two varieties.

Joint probability: The probability of two events occurring together or in succession.

Joint regression: In the classical regression model it is assumed that the dependent variate has been a function, often linear, of a set of independent variables. If the linear model has been indequate, attempts have been sometimes made to allow for cross product terms in the independent variables, e.g., if there have been two independent variables x_1 and z_2 the joint regression equation would be :

$$y=(a_1x_1 + a_2x_2 + a_3(x_1x_2)$$

where the third term allows for any interaction between the original variables. Such regression has been sometimes said to be 'joint' but the usage has not seen to be very desirable.

J-shaped distribution: Refers to an extremely asymmetrical frequency distribution having the maximum frequency at the initial (or final) frequency group and a declining or increasing frequency elsewhere, The shape of distribution roughly resembles the letter *i* or its reverse. Among theoretical curves referred to as J-shaped have been the Pareto and certain of the Pearson system of frequency curves.

Judgment sample Refers to any sample which is not a probability sample.

Judgment sampling/purposive sampling: Any method of sampling where Instead of using probability sampling investigator uses his own expert judgement to select elements which he considers to be typical or representative of the population.

Jump matrix: For a Nlatkov chair in continuous time, the matrix whose (i, j) element has been the conditional probability of a jump from state i to state j, given that a jump from state *i* occurs.

Jump statistic: Refers to a statistic which is used to assess the reality of an apparent jump in the mean of a stationary time series.

K

Kagan-Linnik-Rao theorem: This theorem, which gets characterised the normal distribution by the property of constancy of regression of the sample mean on the deviations from this mean of the sample observations, states that if $X_1, X_2, \ldots X_n$, $n \geq 3$, have been independent and identically distributed random variables with finite expectation $\xi X_i = 0$.

Kamat's Test: Refers to a nonparametric test for the equality of the scale parameters of two populations known to be having equal medians.

Karber's method: Refers to a method which is used for estimating the median effective dose of a stimulus from data on quantal responses.

K-Class estimator: In econometric regression analysis, this term refers to a general class of estimators which were proposed by Theil for the parameters in $By+\lambda z=u$ where y are the endogenous variables, z are exogenous and u are random errors. The method proceeds by least squares in stages.

Kendall's coefficient of concordance: Refers to a measure of the level of agreement (concordance) between several sets of matched results. Symbol : W. It takes values between 0 and 1. If there have been only two judges, a value of W near 0 implies disagreement between them; if there have been many, a value near 0 means randomness in their judgments. A value near I reveals agreement between the judges.

Kendall's 'S' score: In rank correlation analysis, Kendall defined a score for a pair of rankings of n items, as + 1 if any two have been ranked in the game order by the two rankings, -1 if in opposite order and zero if tied in either or both rankings. The total score S has been the algebraic sum of the $\frac{1}{2}n\ (n-1)$ contributions from pairs of items.

Kendall' tau (τ)**:** Refers to a coefficient of rank correlation which is based on the number of inversions in one ranking as compared with another.

Kendall's terminology: Refers to a system of describing particular cases in queueing theory which was developed by D.G. Kendall. It has been based upon initial letters of key words associated with the input, queue discipline and service mechanism of a congestion situation. For example, a random, Poissonian or Markov input may be represented by M and the number of service channels as 1 or s (= 1, 2, ...)

Kesten's process: Refers to a variation of the Robbins-Munro stochastic approximation process which is intended to reduce the bias related to the starting point. The step constant gets reduced only when there has been evidence that doses are straddling the median lethal dose.

Kollectiv: This word denotes a population of objects in which each unit bears one of a finite number of identifiable characteristics.

Kolmogorov axioms: Refers to a set of axioms which were given by Kolmogorov for the foundation of probability theory in terms of set and measure theory. They have been forming the starting point of most modern expositions of a mathematical theory of probability.

Kolmogorov representation: Refers to a representation of the logarithm of the characteristic function θ(t) of infinitely divisible distributions having finite second moment. The representation is having the form

$$\log\ \theta(t) = i\beta t + \int \left[\frac{e^{itu} - 1 - itu}{u^2}\right] dK(u)$$

where β is a real constant, and the function K coincides, to within a non-negative multiplicative constant, with a distribution function.

Kolmogorov-Smirnov Distributions: Refer to a name assigned to a variety of distributions which have been exact or approximate distributions of functions of the sample cumulative distribution function.

Kolmogorov-Smirnov test: Is a non-parametric test, which does not require that data be grouped in any way, for determining whether there is a significant difference between an observed frequency distribution and a theoretical frequency distribution.

Kolmogorov theorem: Refer to two theorems which were stated by Kolmogorov and give the conditions under which the strong law of large numbers holds :

(i) if the random variables are independent

(ii) if the random variables are independent and identically distributed.

Konyus index number: Refers to the name which is assigned to a class of index number rather than to one specific formula. It is an index of prices based on quantities, the 'budget' of the consumer, which are optimal in some field of consumer preference. If the indifference level for consumers has been specified in terms of an optimal budget at the base prices, the index is termed as Laspeyres-Konyus; if by an optimal budget at prices in the period under comparison with the base period, a Paasche- Konyus index.

Kronecker product of design: It was concluded that an experiment design D uniquely as certain its incidence matrix and vice versa. Thus the Kronecker product (in the sense of matrix theory) of two incidence matrices produces a new design matrix and can be used to as certain the existence of certain classes of design

Kronecker product of matrices: The product A x B of an m x m matrix A and an *n* x *n* matrix B has been the *mn* x *mn* matrix whose elements have been the products of terms, one from A and one from B.

Kruskal-Wallis one-way analysis of variance: A nonparametric test of the null hypothesis that three or more samples are drawn from the same parent population. Symbol H. The test has been performed on the overall ranks of the items sampled, and so has been a test for differences in location of the particular samples.

Kruskal-Wallis test: Refers to a multi-sample rank-randomisation test for identical populations, sensitive to unequal locations. It has been a direct generalisation of the two- sample Wilcoxon rank sum test to the multisample case.

K-samples problem: Refers to the problem of determining whether, given *k* samples, one from each of *k* populations, the parent populations have been different. The usual tests developed in this connection have been homogeneity tests for means or variances, but an infinite number of tests have been Possible.

K-statistics: Refers to a set Of symmetric functions of sample values which were proposed by R. A Fisher the univariate k-statistic of order *r* may be defined as the statistic whose mean value has been the rth cumulant, k_r, of the parent Population. The statistics have semi-invariant properties and their sampling cumulants can be got directly by combinatorial methods.

Similarly the multivariate k-statistic, say $k(r, s, \ldots, v)$ may be defined as the symmetric function of observations whose mean value is the corresponding cumulant $k_{rs}, \ldots$ It is more usually written $K_{rs \ldots v}$

Another gereralisation has been due to Tukey, who defines a statistic say $k\{r, s, ..., v\}$ as the symmetric function whose value is $k_1, k_s .....k_v$. This statistic has been, also often written $k_{,,}$ and should not get confused with the multi variate k-statistic.

Kurtosis: The sharpness of a peak on a curve of a probability density function.

A distribution with high kurtosis is termed as *leptokurtic* ; one with medium kurtosis, *meso-kurtic,* and one with low kurtosis, platykurtic.

A measure of kurtosis has been the *coefficient of kurtosis.*

If a curve is sharply peaked at the centre, the curve is leptokurtic. If the central section of the curve is flattened or rounded it is platykurtic.

An intermediate degree of peakedness such as that of the normal distribution is called mesokurtic.

A curve showing the variation in the kurtosis of one variate against values of the other in bivariate variation is called a kurtic curve.

Kuder Richardson Formula: A formula which is used for estimating the reliability coefficient of a test which has been endeavoured to overcome the disadvantages of formulae associated with split half methods. There have been numerous versions of the formula, which may be written

$$r_{tt} = \left(\frac{n}{n-1}\right)\left(\frac{\sigma_1^2 - \Sigma pq}{\sigma_t^2}\right)$$

where n denotes the number of items in the test, p the proportion passing or satisfactorily responding to an item, $q = 1 - p$ and $\sigma_t 2$ the overall variance of scores.

Kuiper statistic: It is a goodness of fit test statistic which is similar to the Kolmogorov-Smirnov statistic. For the single-sample problem it is

$V_n = n½ \; [Sup_x\{F_n(x)\text{-}F(x)\}\text{-}Inf_x\{F_n)\text{-}F(x)\}]$ and for the two-sample problem.

L

Labour-time function: Man hours or man days expressed as a function of work done usually of linear form, and not to be confused with the employment function, which is of logarithmic form.

Labour turnover: The ratio of the number of personnel leaving the firm, in unit time, to the total number employed. Unit time is usually a calendar month, and the-number employed is the average during the month, the average being, say, that of the number employed on the first and last working days and the working day nearest the middle of the month.

Ladder indices: A ladder index may be defined as the time point in a random walk at which a ladder variable begins to take place.

Ladder variable: If, in a random walk, the sum of the mutually independent variables at time point n is Sn then, if $S_n > S_j$ (j = 0, 1, 2, ..., n - 1), the variable X,, has been an ascending ladder variable. Descending ladder variables are similarly defined where $S_n < S_j$.

Lag: An event occurring at time $t + k$ ($k>0$) is considered to lag behind event occurring at time t, the extent of the lag being k. An event occurring k time units before another may be considered as having a negative lag.

Lag correlation: Refers to the correlation between two series where one of the series is having a lag with reference to the other.

Lag hysteresis: Lag hysteresis in econometrics has been referred to cases of sinusoidal variation in the two variables : skew hysteresis is referred to the more realistic case in which the oscillatory movements are asymmetrical. The value of the concept where oscillatory movements have been not cyclic has been unknown.

Lag regression: Refers to a regression in which the values of the dependent variate and one at least of the independent variable get lagged in relation to each other.

Laissez-faire: A doctrine that the economic affairs of society are best guided by the decisions of individuals to the virtual exclusion of collective authority. The idea has its basis in the writings of the Physiocrats, but its analytic foundations lie in the work of Adam Smith and the classical school.

Adam Smith argued that individuals acting purely out of self interest will be a progressive force for the maximization of the total wealth of a nation. The role of the authorities, given this aim, should be permissive, creating a legal defensive apparatus sufficient to allow individual action. Interference with the free working of this natural order will reduce the growth of wealth and misdirect resources.

The theme was extended by the neoclassical writers to include not just freedom of production and accumulation but also freedom of choice in consumption. The model of perfect competition provided further analytic evidence that free competition and unrestricted expression of consumer preferences maximize the economic welfare of society. The doctrine in this form presupposes that collective action will be beneficial if in addition to providing the basic requirements of legal property ownership and national defence it will also act to maintain competition by breaking up monopoly and outlawing noncompetitive practices.

The doctrine has been much criticized for its ideological overtones and especially its implication for economic policy, The apparent inability of a free market economy in the twentieth century to achieve many important social goals such as full employment and stable growth has led to an abandonment of laissez-faire economic policy.

The Keynesian Revolution, problems of international trade, and a concern for income distribution have caused many economists to urge the use of some degree of state planning in order to ensure that broader social aims are met. On the other hand an increasing number of economists stresses the advantages of a competitive economy based on private property and unrestricted industrial action in all spheres of economic activity.

Lancaster's partition of chisquares: A total chi-square value has been computed to provide a gross measure of the extent to which the cell frequencies get departed from expectation. This value has been partitioned into additive components to show how much has been attributable to individual classifications and how much is due to both first order and second order inferaction. The procedure can be extended to cover contingency table involving more than three way of classification where higher order interactions occur. A special merit of this procedure -has been that it can be used with theoretical parameters or with parameters estimated from the data.

Large numbers, law of: A general form of this fundamental law relating to random variables could be stated as follows:

If x_k denotes a sequence of mutually independent variates with a common distribution and if the expectation $\mu = \xi(x_k)$ exists, then for every $\varepsilon > 0$ as $n \to \infty$ the probability

$$\Pr\left(\left|\frac{X_1 + \ldots + X_n}{n} - \mu\right| > \epsilon\right) \to 0.$$

Laspeyres' index: A form of index number due to Laspeyres. If the prices of a set of commodities in abase period are p_o, p_o', p_o'',...and those in a given period p_n, p_n', p_n'',...; and if q_o, q_o', q_o'' are the quantities sold in the base period, the Laspeyres' price index number is written

$$I_{on} = \frac{\Sigma\,(p_n q_o)}{\Sigma\,(p_o q_o)}$$

where the summation takes place over commodities. In short, the prices are weighted by quantities in the base period.

Generally, an index number of the above form is said to be of the Laspeyres' type even when p and q do not relate to prices or quantities; the characteristic feature being that the weights relate to the base period, as contrasted with the Paasche index in which they relate to the given period.

Lesile's test: In capture-recapture experiments, the probabilities of capturing each living member of the population must be the same. In the order to test this assumption, Leslie proposed the statistic

$$L = \sum_{i-1}^{N} (R_i - \bar{R})^2 \Big/ \sum_{g=f}^{g} p_j q$$

where R_i = the number of recaptures of individual i between the fth and gth sample inclusive;

$= \bar{R} = \sum_{i=1}^{n} R_i / N$ and P_j is the proportion of recaptures at the jth sample. The statistic has been distributed as χ^2 with N–1 d.f., although it was later shown by Carothers that (1- 1/N) L has been identical to Cochran's Q.test.

Laspeyres method: To weight an aggregates index, this method uses as weights the quantities consumed during the base period.

Latent structure: In general this term refers to a structure expressed in terms of variates or variables which have been 'latent' in the sense of not being directly observable. Certain econometric relations e.g., in terms of 'utility' have been of this type, and the models used in factor analysis as used in psychology also may be considered as a kind of latent structure.

Latent variable: A variable which has been unobservable but has been Supposed to enter into the structure of a system under study, such as demand in economics or the 'general' factor in psychology, Unobservable quantities such as errors have been not usually described as latent.

Latin cube: Refers to an extension of the principle of the latin square so that the layers, and sections in two directions, have been effectively latin squares.

Latin rectangle: An experimental design which has been derived from the latin square. It consists of a latin square with one or more adjacent rows or columns added or omitted. This particular design has been one form of the incomplete latin square or Youden square.

Latin square, latin square design: A term used in the design of experiments for a design which attempts to remove the experimental error resulting From the—variation from two sources which can be identified with the rows and columns of the square by permutation of the treatments, each of which is identified with a Latin letter.

Lattice diagram: In quality control, it is a diagram on which the results of attribute tests-on samples are illustrated.

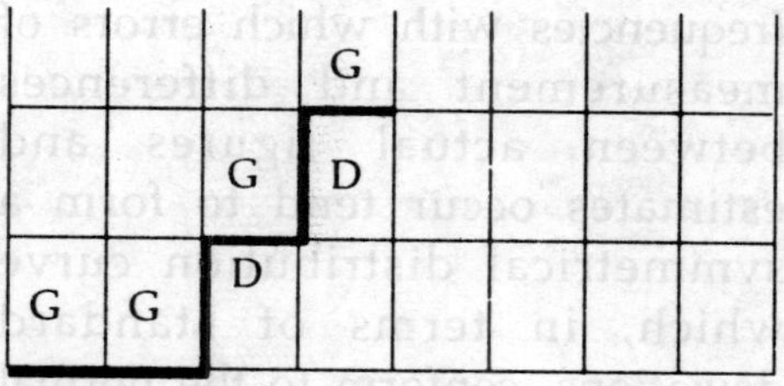

Fig. 1. Lattice diagram. This sampling line illustrates whether the first six items to be tested were good G or defective D ina quality-control test.

Lattice distributions: Refer to a class of distribution, which includes most of the discrete distributions used in statistics, in which the intervals between any one random variable, for which there have been non zero probabilities, have been integral multiples of one quantity and thus form *n*, lattice. By suitable transformations all variables could be made to take values which are integers.

Lattice sampling: Refers to a method of sampling in which Substrata have been selected, for the sampling of individuals, according to some pattern analogous to the allocation of treatments on a lattice experimental design.

Laurent process: A stochastic process for which the covariance generating function has been a Laurent series; the important, case has been that in which the process has been generated by a moving average of a random process.

Law of error: States that the frequencies with which errors of measurement and differences between actual figures and estimates occur tend to form a symmetrical distribution curve which, in terms of standard deviations, conform to the normal curve error.

Law of large numbers: The law is a popular interpretation of the principle of statistical regularity and states that groups or aggregates of data show a greater degree of stability than small ones. Thus if a reasonably large sample or items is drawn at random from a parent group it will probably show the characteristics of the Parent group more representatively than a small sample would do, In fact, the sampling error is inversely proportionate of the root of the number of items in the sample.

Learning curve: A curve which describes the upward trend in productivity , or the downward trend in labour requirements, per unit of output as the employees gain experience in carrying out particular processes or in producing certain goods.

Least favourable distribution. Given the composite hypothesis H_0: f_θ $\theta \in \omega$ to be tested against the simple alternative H, the general procedure has been for reduce the composite hypothesis; H_0 to a simple one $H_0\lambda$ where l is a probability distribution function over ω The maximum power β_2 that can be obtained against H1 has been that of the most powerful test of $H\theta\lambda$ against H_1. The probability distribution λ has been said to be least favourable for a given level a if, for all λ the inequality $\beta\lambda < \beta\lambda'$ holds.

Least significant difference test: Refrts to a test for comparing mean values arising in analysis of variance. It has been an extension of the standard t-test for the difference between two specified mean values, As the tests between pairs are not independent, the error rate is difficult to assess exactly.

Least squares estimate: A estimate obtained by a least squares process or method. For example, in least squares regression analysis a line of be fit is obtained and the relationship between observations of a dependent variable and the regressors is expressed as an equation. The exact point on the line corresponding to the observations estimate not in the usual statistical sense but in the more' general sense that they are 'values produced by a rule of estimation.

Least squares estimator: In statistical analysis an estimator is a rule by means of which the parameters of a parent group may be estimated, the values produced being termed estimates. In any technique using least squares method, the estimator is the rule or model obtained by least squares method; and the estimate is the value produced by the rule.

Least squares method: A method of estimating a function or line of best fit by employing the principle that the optimal functions test which minimizes he squares of differences between the estimates which it produces and the observations on which it is based.

Least variance difference method: It is a method of generalised linear estimation of structural parameters.

Leslie's test, In capture- recapture experiments, the probability of capturing each living member of the population must be same.

Level map: A graph which shows curves in an (x, y) plane corresponding to various values of the constant k in a defining equation $f(x, y) = k$. It has been similar to the contours of equal height on a geographical map.

Level of measurement: A term used to denote the strength of the scale of measurement on which the data are measured. The lowest level of measurement is the nominal scale for which the only applicable statistical methods are those based on frequencies or counting the made, contingency tables, etc.

Level of signifiance: There are many statistical tests of hypothesis which depend or the use of the probability distributions of a statistic chosen for the purpose of the particular test. When the hypothesis has been true this distribution has a known form, at least approximately, and the probability $Pr(t \geq t_1)$ or $Pr(t < t_0)$ can be determined for assigned t_0, or t_1. The acceptability of the hypothesis has been usually discussed, inter alia, in terms of the values of f observed : if they have a small probability, in the sense of falling outside the range to to t_1, $Pr(t \geq t_1)$ and $Pr(t \leq t_0)$ small, the hypothesis is rejected. The probabilities $Pr(t \geq t_1)$ and $Pr(t \leq t_0)$ are called levels of significance and are usually expressed as percentages, e.g 5 per cent. The actual values are, of course, arbitrary., but popular values are 5, 1 and 0.1 per cent.

Levy-khinchine representation: Refers to a representation of the logarithm of the characteristic function $\phi(t)$ of an arbitrary infinitely divisible distribution.

Levy-pareto distribution: The work of Levy in connection with stable distributions proved that the Pareto distribution could be derived from a version of the central limit theorem in which the individual random variables do not have finite variance.

Levy representation: Refers to a representation of the logarithm of the characteristic function $\theta(t)$ of an arbitrary infinitely divisible distribution. The representation is having the form

$$\log \theta(t)=i\beta t-\sigma^2t^2 | 2+\int_0^{\infty} h(t,u)\, dN(u)$$

$$+\int_{-\infty}^{0} h(t, u)\, dM(u)$$

where β is real, $\sigma>0$, $h(t, u)=e^{itu}-1-itu/(1 + u^3)$ and the functions M and N are able to satisfy a number of conditions.

Lexis ratio: This ratio is regarded as a measure of distinguishing the three kinds of variation in sampling for attributes Beraoullian, Lexian and Poissonian.

Liapounov's inequality: An inequality which had been due to Liapounov. It is concerning the relations between the absolute moments of a frequency distribution.

Liapounov's theorem: Refers to a form of the central limit theorem which assumes the existence of absolute third moments.

Life table: A table which shows the number of persons who, of a given number born or living at a specified age, live to attain successive higher ages, together with the numbers who die in the intervals.

Likelihood: A measure of the plausibility of a postulated population structure given values of observations on a random sample from that population. The postulated population; structure is usually in terms of a model for the distribution function of a random variable defined on that population together with values for parameters of that function.

Likelihood function: This is likelihood which is expressed as an algebraic function of possible values of parameters for a particular population distribution and the observed set of sample data.

Likelihood ratio: Refers to the ratio of the likelihood of one postulated population structure to another for a given set of sample data. Likelihood ratios form the basis for likelihood ratio tests.

If $x_1, x_2, ...x_n$ are a random sample from a population $f(x; \theta_1,\theta_2,\theta_\kappa)$ the likelihood of this particular sample would be

$$L = \prod_{i=1}^{n} f(x_i, \theta_1, \theta_2,...., \theta_k)$$

Likelihood ratio tests: These have been tests of null hypothesis of a Population structure against specified alternative hypotheses which are based on the likelihood ratio. Usually the null hypothesis is simple in that it specific the population distribution and values of parameters. The likelihood of th null hypothesis given the sample data can be known by substituting these values in the likelihood function, The alter native hypothesis may sometimes be simple in that it specific particular values for, the parameters.

Limited information methods: In econometrics, these refer to methods of deriving estimates of parameters in a stochastic system which are not using all the information available. The term's generally confined to those methods which give consistent estimates, i.e., are unbiassed for large samples.

Lincoln index: It is a method of estimating the size of population of mobile units, e.g., animal populations which relies on the capture, marking, release and recapture of samples from the Population under investigation.

Lindeberg-filler-theorem: Refers to a form of the central limit theorem which provides a necessary and sufficient condition for the distribution of a sum of independent random variables to be asymptotically normal:

Lindeberg-levy theorem: Refers to a particular case of the central limit theorem when all the variates concerned are having the same distribution.

Lindley's theorem: Refers to a limit theorem in queueing theory which concerns the distribution function of the waiting time of the nth arriving customer.

Linear constraint: Refers to a condition imposed on certain variate values or frequencies which has been linear in form.

Linear correlation: An absolute expression. It was used to denote either (a) the product moment correlation in cases where the corresponding regressions were linear or (b) a coefficient of correlation constructed from linear functions of the observations. The expression has been best avoided altogether.

Linear discriminant function: Refers to a discriminant function which is a linear function of observed variate.

Linear estimator: An estimator which has been a linear function of the sample values. A linear estimator for the parent mean has been:

$$\frac{a_1 x_1 + a_2 x_2 + \ldots + a_n x_n}{a_1 + a_2 + \ldots + a_n}$$

where $x_1, x_2, \ldots, x_n$ have been the sample values and $a_1, a_2, \ldots, a_n$ all constants.

In the case when $a_1 = a_2 = a_n = 1$, this becomes

$$\frac{x_1 + x_2 + \ldots + x_n}{n}$$

where is the sample mean, x.

Linear hypothesis: Logically, this expression is used to relate to any statistical hypothesis concerning the parameters of a distribution which could be expressed linearly in terms of them.

Linear maximum likelihood method: Refers to a method of parameter estimation in multi-equation model in which the Taylor series linear approximations of the partial derivatives of the likelihood function get equated to zero rather than the actual partial derivatives.

Linear model: Refers to a model in which the equations connecting the variates or variables have been in a linear from.

Link relative: In index number theory, the value of a magnitude in a given period divided by the value in the previous period. It is trend ratio.

Line sampling: Refers to a method of sampling in a geographical area. Lines have been drawn across the area and all members of the population falling on the line, or intersected by it. If the lines have been straight parallels equally spaced across the area concerned, then the sampling one form of systematic sampling. If, instead, of all intercepts of the lines, a series of evenly spaced points have been chosen each line, the sampling has been equivalent to choosing the points on a lattice and may also be considered as two stage line sampling.

Linked samples: Two samples of the same size in which there is a one to one correspondence between their respective sample units. The linking method may provide either an absolute of partial constraint on the selection insofar as the selection of a unit in a sample may determine uniquely the element to be selected in the other, or may simply restrict the choice of the corresponding element.

List sample: Refers to a sample which is selected by taking entries from a list of the items constituting the population under review. The usual method of selecting entries has been to take them at equal intervals, the starting point being selected at random.

Local statistic: Refers to a statistic, estimator or test statistic, which gets derived from short term comparisons within a time series.

Location parameter: Refers to a parameter which locates a frequency distribution in the sense of defining a central or a typical value like a mean or mode.

Logarithmic chart: A graph where on one or both axes have been scaled in terms of logarithmic of the variables. The chart may be termed as a semi or double logarithmic chart according to whether only the ordinate or both the ordinate and abscissa are on a logarithmic scale.

Logarithmic normal distribution: If the logarithms of a set, of variate values have been distributed according to the normal distribution the variate has been said to have a logarithmic normal distribution, or be distributed 'lognormally'.

Logarithmic scale: A chart, which is drawn to logarithmic scale in one or both of its axes uses intervals and gradations of equivalent logarithm distances instead of normal ordinate and co-ordinate measurements.

Logarithmic distribution: Refers to a frequency distribution of the form $\beta\{e^t/(1+e^t)^2\}$ with $p>0$. It closely approximates the normal distribution and could also be shown to be the asymptotic distribution of the mid-range of an exponential-type parent type distribution. In its cumulative form it is also used as a growth curve.

Logit: In some problems relating to the proportion of subjects responding to different doses of a stimulus the model based on a tolerance distribution has been not appropriate. A logistic relation

$$P = \{1+e^{-(a+\beta x)}\}^{-1}$$

may approximately denote the dependence of probability of response P on dose x

Lorenz curve: A graph which has been designed to show the extent of inequality in the distribution of a variable throughout a population. Perecentiles of the population have been plotted against percentiles of the variable concerned.

Loss function: In the making of decision on the basis of the observation through ignorance of the true distribution of x. The extent of the disadvantage has been often a function of the true distribution and of the decision which is actually made. This function has been called the loss function.

Lottery sampling: Refers to a method of drawing radom samples from a population by constructing a miniature of the Population, e.g., by inscribing the particulars of each member on to a card and drawing members at random from it, e g,, by shuffling the cards and dealing a set a set haphazardly.

Lower-tailed test: A one tailed hypothesis test in which a sample value Significantly below the hypothesized population value will lead us to reject the null hypothesis.

L-statistics: Refers to a class of rank order statistics which was proposed by Dwass in connection with rank order tests in the two sample problem; this must not get confused with the Neyman & Pearson L-test for homogeneity of sample variance,. The L-statistic may be defined as $\Sigma aN_i bNR_i$ where the aN_i have been $(n/mN)^{1/2}$-i = 1,. . .,m, and i = m+l,...,N being the two sample under *t*est--the *bN* another set of constants and $R_1...R_n$ the ranks of the N random variables.

M

Magic square design: Refers to a square array ($n \times n$) with n^2 integers kept in the cells in such a way that the row column and two principal diagonal sums have been the same namely. ¼ n (n^2+1) These squares may be employed for balancing out linear trend from main effects and lower order interactions in, some factorial designs and in some Latin and Groeco Latin square designs.

Malthus' theory: According to this theory the human population will expand exponentially while the means of subsistence will increase more slowly, with the consequence that ultimately Its size will be kept in check by famine, disease and war. It was given by the English economist T.R. Malthus The widespread use of birth control methods during this century has made much smaller rates of population growth possible: consequently the theory has been now mainly of historical interest.

Manifold classification: If you decide a population into a number of mutually exclusive classes according to some given characteristic and then each class gets divided by reference some second third etc. characteristic, the final grouping is termed as a manifold classification.

Mann-whitney u-test: Refers to a test of significance which is applicable to two independent samples where measurement is on at lest the ordinal level. The test statistic U assesses the probability of a particular pattern of results in relation to the total number of possible patterns. The test regards that total number of possible patterns the two samples have been independent and that sampling has been random. The hypothesis being tested is that there is no difference between the from which the sample were drawn.

Marcinkiewicz's theorem: According to this theorem characteristic function ϕ is having the form exp $P(t)$ where P is a polynomial then P is a quadratic polynomial and ϕ is the characteristic function of a Normal distribution.

Marginal category: Refers to one of the frequency classes of a marginal classification.

Marginal classification: In a bivariate frequency table it is customary to show, as row and column totals, the univariate termed as marginal classification.

For example in marginals the distribution indicated in the marginals row and column

Marginal distribution Refers to the distribution indicated in the marginals row and column

2, 10, 26, 41, 26, 12, 2 and
1, 9, 25, 40, 31, 10, 0
have been marginal distributions of frequencies.

Marginal cost: The extra cost of producing an extra unit of output. In the short run the marginal cost curve slopes upwards due to the operation of the law of diminis*hing returns.* Note also that marginal cost cannot be affected by the level *of fixed costs.* Quite simply, if an extra unit of output is produced fixed costs do not change and hence extra fixed costs must be zero.

Marginal cost of funds schedule: The schedule detailing the true cost of financial capital to the enterprise. In an imperfect capital market the true cost will exceed the *interest rate.* Firms can raise money to finance investment from many sources and the cost of finance will vary with the source. The least expensive source of finance may be the firms retained profits or its depreciation fund. More expensive will be a *rights issue* or new issue of *equities* or borrowing on fixed interest.

Marginal cost pricing: A pricing practice pursued by private firms or public corporations in which price is made equal to *marginal cost.* Given continuous revenue and cost curves, this implies setting price at the point at which the demand curve cuts the marginal cost curve. The market conditions prevailing in perfect competition ensure marginal cost pricing since average and marginal revenue are the same. Hence the requirement for profit maximization, that marginal cost be equal to the marginal revenue, means that price equals marginal cost. Under *imperfect competition,* however, profits will not be maximized with price equal to marginal cost since average revenue will exceed marginal revenue. Hence marginal cost pricing under imperfect competition will only come about through some form of regulation or taxation. In the public *sector, nationalized industries* are recommended to use marginal cost pricing, the rationale being that it maximizes economic welfare.

Marginal efficiency of capital: That unique *rate of discount* which would make the *present value* of the expecied net returns from a *capital asset just* equal to its supply price when there is no rlse in the supply price *of* the asset. The term originates with *Keynes* and is som, enmes known incorrectly as the *internal rate of return.*

Marginal efficiency of investment: That *rate of discount* which would make the present value of the expected net returns from the *capital asset* just equal to its supply price where it is recognized th^t this price will rise in the short-run.

Marginal efficiency of investment schedule: The demand curve for investment. The schedule detailing the relationship between the *marginal efficiency of investment* and the interest rate. Investment will be carried on until this point in the short run while in the long run the equilibrium of the capital stock is given by the point of equality of the *marginal efficiency of capital* and the interest rate. In contrast to the *marginal efficiency of capital schedule* this schedule describes a flow and not a *stock* relationship.

Marginal per capita reinvestment quotient criterion: An investment criterion with the objective of maximizing income per head at a future date. It states that the best allocation of resources will be achieved by equating the marginal per capita reinvestment quotient of capital in each of its uses. It is assumed that profits are reinvested and wages consumed. Gross productivity per worker minus consumption per worker determines the gross amount per worker available for reinvestment, The criterion thus favours projects where profits are high. The development objective underlying the criterion has been criticized as being unrealistic. Governments are likely to wish consumption to increase in the near future as a main objective. The use of the criterion will not give priority to increasing employment which may well be considered a main objective.

Marginal physical product: This is defined as the addition to total output resulting from the employment of an additional unit of labour, and may be derived from the *production junction* holding other factors fixed. The marginal physical product schedule may be regarded as the demand curve for labour. Labour is employed up to the point at which the payment to the last unit equal to the output produced by the last unit. It is more conventional by means of a simple change in scale to convert marginal physical product into value of marginal physical product namely by multiplying the former by product price. A wage rise, for example, will have a negative *substitution effect* as the firm substitutes cheaper capital for the more expensive labour, thereby reducing the quantity of labour demanded. There will also be a negative scale effect as a wage increasc lcads to a higher *marginal cost* of production which in turn reduces the firm's optimal output and hence *derived demand* for labour. The two effects reinforce each other to lead to an unambiguously inverse relationship between wages and the firm's demand for labour. This demand curve is more elastic than the short-run demand,

Marginal product: The extra *output* obtained by employing one extra unit of a given *input* . Thus the term should be qualified with respect to which input is in question--e.g. the marginal product of tabour, the marginal product of capital, etc.

Marginal probability: In a two-way table, showing the probabilities of outcomes of two events, the row and column totals provide the separate probabilities of outcomes of each of the events.

Marginal revenue: Refers to the change in gross revenue due to a change of one unit in the rate of sales resulting from a change in price. Its formula is as follows:

$$M = p\left(1 - \frac{1}{e}\right)$$

where M is the marginal revenue, p the price, and e the price elasticity of demand. The price elasticity of demand is an econometric concept of considerable practical significance.

Marginals: These are the total univariate frequencies of two variates in a variates frequency table, shown separately in the marginal row and marginal column of the table. The term 'marginal' in this sense means 'at the margin' or 'at the end of the tow or column' and must not be confused with the particular sense of 'differential', which the word has in economic analysis.

Markov chain: A series of discontinuous states at intervals of time $(t_1, t_2 ... t_p)$ such that the relative size of constituent classes or groups in a population at any given time is determined by stochastic probabilities inherent in the previous state, and not in any way affected by the earlier state (s) of the system.

Markov estimate: Refers to an estimate of a parameter derived from an estimator giver by the so-called Markov or Gauss-Markov theorem.

Markov inequality: It a variate x has been non-negative and has mean equal to a then for $t > 0$ the

Markov inequality states that

$$\Pr(x \geq t) < \frac{a}{t}$$

Some writers owe this inequality to Tchebychev.

Markov process: Refers to a stochastic process such that the conditional probability distribution for the state at an future instant, given the present state, has been unaffected by any additional knowledge of the past history of the system,

Marshall edgeworth-bowley index: It is an index number formula which was proposed by Marshall, Edgeworth and Bowley as an alternative to the standard formulae of Laspeyres and Paasche. It may be written, in terms of prices and quantities, as:

$$I_{on} = \frac{\Sigma\{p_n(q_o + q_n)\}}{\Sigma\{p_o(q_o + q_n)\}}$$

and where the suffix *o* refers to the base period and π to the period to which the index relates. The index is a compromise with no general bias.

Marshall-Olkin distribution: Refers to a multivariate exponential distribution which was introduced by Marshall & Olkin.

Martingale: Originally, this term refers to a process known gamblers under which the loser at a fair game doubled his stakes for the next, and so on at each loss, the paradox being that in the long run heappeared certain to win sooner or later and at that point would have a net gain. More recently the term has been assigned a precise significance in the theory of stochastic processes.

Master sample: Refers to a sample which is drawn from a population for use on a number of future occasions, so as to avoid *ad hoc* sampling, on each occasion. Sometimes the Master Sample has been large and subsequent inquiries have been based on a sub-sample from it.

Matched groups: Refer to groups which are matched for similarity before an experiment is carried out on one of them, the other being used is the control group,

Matched samples If the elements in a pair (or set of samples are such that each element in one sample corresponds to a particular clement in the other on some characteristic other than the variable being investigated, the samples are said to be matched.

Matching: If we compare the two sequences of a finite number of characteristics A_1, A_2.., A_k the jth of one sequence against the jth of the other a comparison in which both members, exhibit the same characteristic is termed as a match. The number of' matches in an observed pair of sequences gives a test of various hypotheses concerning the system which generated them.

Matching distribution: Suppose two sets having it objects each have been numbered 1, 2,...,n and each has been arranged in a random order so as to form n pairs. Then the number of pairs on which the two numbers have been the same has the matching distribution.

Matrix sampling: The process of taking a bi-sample from an R × C population matrix whose elements are $||x_{ij}||$ has two samples, r of rows and c of columns, and forming a matrix whose elements have been those at the intersections of the selected rows and columns.

Maverick: A term used in industrial statistics to denote an observation lying so far outside the usual range that it has been suspected of not belonging to the population under inquiry.

Maximum likelihood estimators: It is a method of estimating the parameters of a population model from the sample data through the likelihood function.

Maximum likelihood method: It is a method or estimating a parameter or parameters of a population by that value or values) which maximises (or maximise) the likelihood of a sample.

Maxwell distribution: Refers to a fortunately rate expression for a chi-squared distribution having three degrees of freedom It is based presumably on Clark Maxwell's discussion of the energy of particles moving at random in three dimensions.

Median voter theorem: A theorem concerning collective choice in a democratic society which predicis that poloiticians will most closely represent the views of those voters at the centre of political or social spectrum.

Mean: The most commonly used average or measure or central tendency or measure of location. The mean is computed by adding together the observations in a series and dividing by the number of observations, in the series.

Mean absolute deviation: Refers to the mean of the absolute values of the differences between the values of a variable and the mean of its distribution.

Mean absolute error: Refers to an alternative but much less preferable name for the mean deviation.

Mean deviation: A measure of the spread or dispersion of a sample of population. Deviations of individual observations from a reference point, usually the arithmetic mean, are calculated. Negative signs are ignored so that only magnitudes of deviations are recorded and not their directions. These deviations are then averaged using an arithmetic mean. Thus if $x_1, x_2...x_n$ are observations and m is their mean (or reference point)

$$MD = \frac{1}{n}\sum_{i=1}^{n} |x_i - m|$$

For grouped data where a frequency f_1 is associated with x_1

$$MD = \frac{1}{n}\sum_{i=1}^{k} f_i\ |x_i - m|$$

Where $N = \sum_{i=1}^{k} f_i$

Mean deviation, coefficient of: Mean deviation expressed as a proportion of the mean. The coefficient is useful for comparing the mean deviations of distributions with dissimilar means.

Mean difference: Refers to the average of the absolute differences of all possible pairs of values of a variate.

Mean profit difference: Refers to a measure of the difference between two series of observations giving parallel profit regression lines.

Mean range: Refers to the, arithmetic mean of the ranges of a set of samples of the same size. The mean range in repeated sampling can be used as an estimator for the population standard deviation.

Mean range method: It is a rapid method of approximating the standard deviation of a population from the ranges of samples. It involves: (1) the calculation of the mean range of the samples; (2) the multiplication on the mean range by a factor whose value is determined by sample size. These factors may obtained from a statistical table, an extract of which is given below.

Mean square: Refers to the arithmetic mean of the squares differences between a set of given observed values of a variate and a particular value of the variate. In a restricted sense the variance of set of values has been a mean square, with particular reference to the mean of the observed values of the variate The term mean of the square has been more general and may be understood as the mean of squares of differences between observed values and any fixed value. The term is having a slightly different meaning in anal of variance. In compiling an analysis of variance table, sums of squares for the variables are determined. The sum squares are then divided by the appropriate number of degrees of freedom.

Mean square deviation/mean square error: Both these terms have been strictly understood in much the same sense as mean square, i.e., the average of the squares of differences between each of a set of observed values of a variate and a given value. If the *given value* used as a term of reference has been the mean, then all three terms are identical with variance.

Mean successive difference: In a time series, this term refers to arithmetic mean of the difference of successive values, *i.e.,* for a series x_1, $x_2 \ldots x_n$. is

$$d = \frac{1}{n-1} \sum_{i=1}^{n-1} |x_i - x_i + 1|$$

Mean values: Mean values have been a general class of functions of distributions of which moments constitute a special case. If a variate is having a distribution function F(x) and t(x) be some function defined within the range of the distribution, the mean value, or mathematical expectation, may be defined as

$$\xi[t(x)] = \int_{-\infty}^{\infty} t(x)dF(x),$$

subject, of course, to existence. The definition could readily be generalised, to n-dimensional -variation. The various moments of a distribution have been derived by substituting the appropriate function for, $t(x)$.

Measures, measurement: Both terms may be used to mean a standard common to a number of objects, by which they may be compared (e.g. a centimetre in measuring distance). The term measurement has more extensive uses, which include :

1. systems of measurement (e.g., pound or dollar for measuring money, or centigrade and absolute scales for measuring temperature); and

2. levels of measurement (e.g., cardinal, ordinal, etc.).

Measures of central tendency: Typical or central values of a variable. Common' used measures of central tendency have been mean (or average), mode median and midrange.

Measure of location: A quality which is able to purport to locate a distribution, or a set of sample values derived there from, by means of a value which is in some, sense central or typical, e.g., the arithmetic means, the median or the model.

Median: The median may be defined as that value of the variate which divides the total frequency into two halves.

Median class: Refers to the class in a frequency distribution that contains the median value for a data set.

Median f-statistic: If a sample of $2m + 1$ items has been ranked and transformed to a uniform distribution the normalised random variable

$$M_{2m+1} = 2 \sqrt{(2m+3)} \{F(x_{m+1}) - ½\}$$

converges in distribution to the standardised Normal variable.

Median lethal dose: Refers to a particular name for the median effective dose when the response has been death. It is often written LD_{50}.

Median line: Refers to any line in a plane which divides a even set of points into two equally numerous sets.

Median regression curve: Refers to the type of regression line or curve which is derived from the Mood-Brown procedure.

Median test: A rank-order test which was proposed by Westenberg (1948) and Mood (1950). It rejects the hypothesis of identity of two populations, in the one sided case. When there are too few observations from one sample larger or smaller than the median of the combined sample.

Median test is a non-Parametric test of similarity between two populations which relates the median of the two samples to the median of the combined sample.

Median unbiassed confidence interval: It is a (1—α) confidence interval which is bounded by two confidence limits θ and θ″ at (1—α)/2.

Merchant banks: In the UK denotes one of a group of institutions that carry on a very varied range of financial or finance related activities. Amongst these the acceptance of commercial bills and the financing of overseas trade are prominent and the major merchant banks are all accepting houses. The term originated in the last century with institutions that came to combine overseas trade financing with actual merchanting, but which mostly dropped their trading activities to specialize in finance and especially acceptance business. Later, arising from their overseas connections, the larger houses became prominently involved in raising long-term finance for foreign governments and large-scale overseas enterprises. In this century this expertise in raising capital has been largely directed to the home market and most merchant banks have become issuing houses, arranging security issues for domestic users of capital. In recent decades they have moved into the broadening area of corporate financial advisory work-assisting in capital reconstructions, advising on mergers and acquisitions--and the similarly expanding business of portfolio management, especially on behalf of pensions funds. As bankers they take deposits mainly in large sums and for defined periods, and lend mostly to corporate borrowers. They are active as foreign exchange dealers, operate extensively in the euro-currency—markets. Different houses combine these activities in differing proportions; some of the smaller ones, financially speaking, retain connections with, or are offshoots of, trading enterprises.

Metameter: Refers to a transformed value, of a dose or a response, e.g., logarithm or profit which is obtained by using at transformation equation that is independent of all parameters. It has been adopted mainly to simplify the analysis or the expression of the dose response relationship.

Mid-mean: Refers to the arithmetic mean of the sample observations which remain after trimming the smallest and the largest 25 per cent of the original sample. If the original sample size has been not a multiple of 4, the two extreme remaining observations after trimming have been appropriately weighted.

Midrange: A measure of location or central tendency. It is the arithmetic mean of the largest and smallest observations in the, data.

Minimal sufficient statistics: A vector of statistics is said to be minimal sufficient if it is having a minimal number of components, in which case it is a function of all other sufficient vectors for the parameters in question.

Minimax estimation: Refers to the estimation of parameters by the application of the minimax principle to a risk function. In particular it is possible to show that a Bayes' estimator which has a constant risk function has been also a minimax estimator.

Minimax strategy: If a strategy gets selected from a group of admissible strategies as being the one which, on a basis of the expected loss, is having the smallest maximum loss this strategy would be a minimax strategy.

Minimum chi-squared: Refers to a method of estimation which is based upon the $\times^2$ goodness of fit statistic. The method determines values of the parameters so as to minimise χ^2 calculated from observed frequencies and 'expected' frequencies expressed in terms of the parameters.

Minimum discrimination information statistic: This statistic has been based upon the principle of information in a sample and can be regarded as the divergence of the alternative hypothesis from the null hypothesis. The distribution of this statistic is involving central χ^2 and it is having additive properties. For a wide class of problems concerning contingency tale, this stastistic takes the form of —2 In λ where λ refers to a likelihood ration.

Minimum variance: When applied to estimates this term is used to represent the property of possessing the least variance among the members of a defined class. A minimum variance estimating exists only where there is existing a sufficient estimator.

Minimum variance linear unbiassed estimator: In the case of the linear- hypothesis the method of least squares gives estimators that have been, unbiassed, linear in the observations, and with minimum variance.

MINQUE: Rao introduced a method called MINQUE (minimum norm quadratic unbiassed estimation) which was used for estimating the variance σ 2 for the Gauss-Markov linear model.

$$Y = X\beta + e$$

Where Y denotes a vector of n observations, X an $n \times m$ known matrix, β a vector of m unknown parameters and e an error vector with a dispersion matrix of diagonal form.

Missing plot/missing value: In an experiment or a survey some planned observations may be missing because of circumstances beyond the control of the researcher. These cases may be given a special code and special techniques are available for analysing data where some of the values are Missing, but these are outside the scope of this book. The term 'plot' arises from the fact that most *such-early*, work was done in the area of agricultural experimentation a missing plot was one for which, for some reason, no observation was available.

Missing plot technique: It is the name assigned by Allen & Wishart and Yates to the process of analysing material which had been designed to conform to an experimental pattern but from which certain values have been missing through circumstances beyond the control of the experimenter. The use of the word 'plot arose from the agricultural back ground of the original investigations noted above but methods have been applicable to missing values generally.

Mitscherlich equations It is possible to estimate the mounts of an element the soil using Mitschelich equation

$$Y = A\ (1\text{-}bR^x)$$

where Y denotes yield and x the amount of fertilise applied

Mixed autoregresive-moving average (ARMA) process: Refers to a stochastic process *{ut}* in discrete time which is driven, by independent and identically distributed random disturbances *{et}* with expectation zero, in accordance to the model.

Mixed autoregressive- regressive systems: Refers to a system of equations which could be able to represent a particular econometric model might be an autoregressive set of x's

$$\sum_{j=1}^{k} a_j y_{t-j} = \sum_{l=1}^{a} \beta_{lt} x_i + \varepsilon_t.$$

Mixed distribution: A term which have been frequently used in, modern statistical literature for the distribution which gets resulted from the parameter of a given distribution being itself a random variable.

Mixed model: This term has been used in at least four slightly different-senses; a model is said to be mixed if

(i) its equations are having both determinate and stochastic' elements;

or

(ii) its equations are having both difference and differential terms

or

(iii) it is having endogenous as well as exogenous elements

or

(iv) in an analysis of variance context for a two way layout the rows correspond to a model factor, i.e., fixed effects and the columns correspond to a model II factor, i.e., random effects. This version is sometimes termed as model III but this is best avoided.

Mixture of distributions: Refer to a process of getting a Weighted average of a group of distribution functions which is a new distribution function. It has to be distinguished from the compound distributions.

Modal group: Occasionally a set of numbers or observations are arranged and summarized as a grouped frequency distribution. The modal group is one with greater frequency for its breadth than either adjacent group. A frequency distribution can have more than one modal group. Modal groups can be defined for non-numeric nominal or ordinal scale frequency distributions.

Mode: The value in a set Of numbers which occurs more frequently than neighbouring Possible values. There may be more than one mode. For example, the mode of ten numbers 1, 1, 2, 3, 3, 3, 7, 8. 9, 9 is 3 - the fifteen numbers 2, 2, 2, 2, 3, 3, 5, 5, 5, 5, 6, 6, 7, 8, 9 have two modes, 2 and 5. The latter set of numbers is said to be bimodal. The normal distribution is unimodal. It has one mode, which is coincident with its mean and median. Asymmetric, or skewed distributions, also have' single modes which do not coincide with their means and medians. In contrast, some frequency distributions are bimodal and multimodal, usually because the frequency distributions are themselves composites of two or more frequency distributions or different categories, which have differing modes.

A rectangular distribution is having no mode.

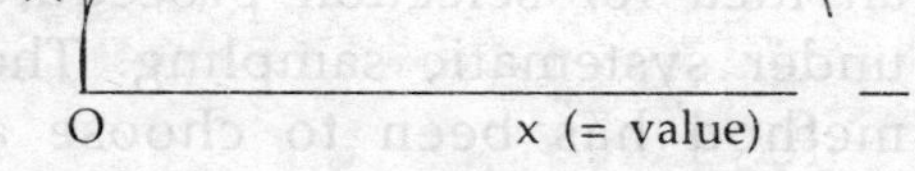

Model: In statistics a model is generally expressed in terms of symbole (i.e., in mathematical form), usually as a set of equations, However, it is possible also to use diagrammatic models but for the purposes of estimation symbolic representation is usually necessary.

Modeling techniques: Methods for deciding which variables to include in a regression model and the different they can be included.

Model sampling: The use of sampling methods which are based on broad assumptions about the way in which the variables of interest are distributed in the population, i.e., they are based population among the methods in this class are haphazard sampling, purposive or judgement sampling and quota sampling all of which are defined elsewhere.

Modified binomial distribution: If it is considered a sequence of n trials with constant probability of success in an individual trail with the restriction that as soon as a success is experienced, the subsequent *m*—1 trials give rise to failures. The number of successes in the above sequence is a modified form of the binomial distribution.

Moving average model: Refers to the representation of a stationary stochastic process in terms of a moving average of infinite length. The observation x_t has been a linear combination of residuals z_t of the form

$$\{x_i\} = \sum_{j=0}^{\infty} b_j z_{j-j}.$$

Moving average process: Refers

to a special case of the moving summation process. If $\{\varepsilon_t\}$ be a random process the process $\{\xi_t\}$ defined by

$$\xi_t = \sum_{i=0}^{k} a_j \varepsilon_{i-j}$$

will exist and be stationary if $\{\varepsilon_t)$ is so. It is termed as a moving average process.

Multiple decision methods: Refers to the statistical techniques which are used for dealing with problems where the possible outcomes or decisions have been more than two.

Multiple decision problem: Refers to the problem of choosing one hypothesis or decision from a set of k mutually exclusive and exhaustive hypotheses or decisions on the basis of some observations on a random variable.

Multiple factor analysis: Now this term is equivalent to factor analysis. Historically, this term refers to the analysis of psychological material into factors grew from one factor to two factor and then to them factor complexes and the latter was termed as multiple to distinguish it from the simpler forms, it no longer seems, necessary.

Multiple markov process: Rafers to a stochastic process in which the transition probabilities have been found to depend on previous values at more than one point. In most contexts the expression has been equivalent to 'autoregressive process'.

Multiple poisson distribution: This distribution has been the joint distribution of independent Poisson variates. It may also be considered as an approximation to the multinomial distribution when in has been large and $np_i = \lambda_i$ in the form:

$$e\text{-}(\lambda_1+\ldots\ldots\lambda_{r-1}) \frac{\lambda_1^{k1}\lambda_2^{k2}\ldots\ldots\lambda_{r-1}^{k_r-1}}{k_1!k_2!\ldots\ldots k_{r-1}!}$$

Multiple poisson process: This term is sometimes used for a general isolation of a Poisson process where there have been simultaneous occurrences at the event points.

Multiple random starts: Refers to an idea for selection procedure under systematic sampling. The method has been to choose a random sample of s without replacement from the first k elements and subsequently every kth multiple element of those selected.

N

Negative binomial distribution: Refers to a distribution in which the relative frequencies (probabilities) have been arrayed by a binomial with *a* negative index. For example, if the variety values have been 0, 1, 2, etc., the frequency at $x = j$ is the coefficient of t in the expansion of $(1\text{-}pt)^{-n}$ $(1\text{-}p)'^{n}$ in powers of *t*. The distribution is sometimes termed as the Pascal distribution.

Negative moments: The negative moments of a frequency distribution may be defined as the moments of reciprocal powers of the variety; for example, $\xi(I/x^k)$ where *k* denotes the order of the moment. Their existence, of course, depends upon convergence and they have been little used. In fact, owing to possible confusion with ordinary moments, which have negative sign, the preferable expression would have been moments of negative order.

Negative multinomial distribution, If there has been a sequence of independent trials in each of which there can be s+1 mutually exclusive outcomes A_u......, A_s having probabilities p_0,...., p_s (Σpi= l) and $x_1,...,x_s$ are the occurrences of A_1.....,A_s before A_0 occurs *k* times, then (x_1,......,A_s (is having the negative multi-nomial distribution.

Nested balanced incomplete block design: Refers to an experiment design having two systems of blocks, the second within the first such that ignoring either system leaves a balanced incomplete design whose blocks have been those of the other system, In the nesting system each block from the first part is having *m* blocks from the second

Nested classification: If one of the factors in a two-way classification has been such that no main effect could be attributed to it, then that factor may be said to be rested within the main classification.

Nested design: Refers to a class of experimental design in which every level of a given factor is appearing with only a single line of any other factor, Factors which are not nested are crossed. If every level of one appears with every level of the others, the factors have been said to be completely crossed: if not, they are partly crossed.

Nested sampling: A term which is used in two somewhat different senses: (1) as equivalent to multi-stage sampling because the higher-stage units get 'nested' in the lower-stage units; (2) where the sampling has been such that certain units get embedded in larger units which form part of the whole sample, e.g., the entry-plots of clusters get 'nested' in this sense.

Net correlation: Refers to an alternative but obsolescent term for partial correlation. The name has been derived from the fact that a partial correlation has been the correlation between two varieties of a larger group when the remaining varieties have been held constant giving, in a sense a 'net' correlation.

Net error: Refers to that part of the total or gross error which does not cancel out when cases or observations have been averaged or totalled.

Net present value: One of the three principal techniques and yield criteria of discounted cash flow; it is equal to the present value of the net yield of a fixed asset over its life less the outlay on the asset. In the discounting process an appropriate current rate of interest is used. The lower the rate of interest used far the purpose, the greater is the calculated present value of the yield. The present value of Rs 1000 a year for 20 years at 15% is Rs 6259; at 10% it is Rs 8514 and at 5%, Rs. 12462. The net yield of the asset should be gross of depreciation, as the NPV technique embodies sinking-fund depreciation provision and also charges interest on capital-both at the rate of interest used in the discounting process.

Net regression coefficient: The coefficient of an independent, Variable in an analysis involving a least one other independent variable; any coefficient in a multiple regression equation.

Net-revenue schedule: A statistical schedule which sets out a range of prices of a brand of product, and the corresponding annual sales in terms of quantity and money, the annual total or variable costs and the profit.

Neutral curve: (curva neutra) Refers to the frequency curve of a distribution having neutral abnormality.

New classical macroeconomics: Essentially a restatement in a more rigorous form of orthodox *classical economics.* This restatement rests on the notion of resources *rational expectations* combined with a natural rate of unemployment which emerges as a result of efficient market clearing, Accordingly it is argued that the economy will settle at the natural rate and attempts to disturb this equilibrium, say to reduce unemployment below the natural rate, will be fully anticipated and neutralized by economic agents. Any attempt to stabilize output or employment below or above the natural rate by fiscal and *monetary policy will* prove ineffectual and will not alter the value of real variables in either the long or the short run. The essence of this view is captured in the proposition that there is no trade-off between inflation and unemployment even in the short run, i.e. that both the short-run *aggregate supply* curve and the short-run Phillip's *curve* are vertical.

New microeconomics: The name given to a body of economic literature which has sought to identify the micro-economics of macroeconomics; specifically, to provide a firm microeconomic rationale of the mechanisms underlying the aggregate relationship between price changes and unemployment. The new microeconomics builds upon the natural rate (of unemployment) hypothesis, and abandons the notion of fully and costlessly informed market participants. Beyond exploring the implications of the natural rate hypothesis (e.g. accelerating inflation where the actual rate of unemployment lies below the natural rate and accelerating deflation, after some point, where unemployment exceeds the natural rate), the new microeconomics seeks to explain the existence of a positive natural rate of unemployment and the negative short-term Phillips Curve relation between wage/price change and unemployment. The new micro-economics models are not identical, but all attribute output changes to price 'surprises'. Thus, unanticipated increases in aggregate demand lead to unanticipated inflation; prices or wages begin rising faster than agents believe they are rising elsewhere in the economy. However, once agents discover the rise in prices and wages to be quite general and not specific to their particular markets, output and employment will return to their previous levels.

'New-new' microeconomics: The name given to model building exercised which attempt to explain wage and price contracts in tērms of microeconomic optimizing behaviour. The existence of wage and price contracts means that prices will react sluggishly to underlying changes in aggregate demand and that the latter will predominantly induce quantity changes in the short run. The casual sequence is from changes in aggregate demand to changes in output and employment which then lead to changes in price and wage inflation.

The new-new micro-economics is commonly contrasted with the new microeconomics, which attributes output or quantity adjustments to price 'surprise'. The new-new micro-economics developed partly in response to perceived limitations of the new microeconornics and in particular, the alleged failure of the latter to provide an explanation of layoffs, here viewed as an involuntary phenomenon) and the counter factual prediction, of a pro-cyclical behaviour of quit. Recent developments have indicated these criticisms to be less damaging than first thought and so the new-new microeconomics is best regarded as an extension of rather than a replacement for the new microeconomics By introducing the institutional features of tabour and product markets it is possible to extend the predictive reach of the latter. Two principal variants of the new-new microeconomics exist in the form of idiosyncratic exchange and implicit contract models.

Neyman allocation: A name given to optimal allocation of a sample when casts are assumed to be equal in each stratum.

Neyman model: This term was given by Ogawa and used to denote experiment designs, which involve situations containing technical errors in the sense that variations in replicate obser-vations have been solely due to experimental techniques.

Neyman-pearson lemma: The Neyman-Pearson lemma provides a methodology for finding the best test among those with the same probability of type I error, i.e., the test with the highest Power (the lowest probability of type II error). The test is based on the likelihood ratio and an example is given in the entry likelihood ratio test.

Nominal scale: The simplest form of measurement has been on a nominal scale. Here individuals or elements get allocated to categories of he characteristic being measured where no relationship is assumed to exits between the categories.

Nomogram: A form of multiple line diagram in which the scales for different functions of a variable are drawn alongside each other and corresponding values are in line with each other. Nomograms facilitate the identifi-cation of differences between scales, but are not as accurate as tables.

Noncentrality parameter: Refers to a parameter of the distribution of a noncentral test statistic that gets vanished in the central case.

Non-central beta distribution: Refers to the distribution of the ratio

$$\frac{\chi'^2 v_1}{\chi^2{}_{v1} + \chi^2{}_{v2}}$$

where χ^2_{v1} has been the noncentral χ^2. If both have been eon-central the resultant distribution has been the doubly non-central beta distribution.

Non-central confidence interval: Refers to a confidence interval which has been not central.

Non-central χ^2 distribution: The distribution of the sum of squares of independent Normal get variated with unit variance but not with zero mean. The distribution was given by R. A. Fisher in 1928 as a special case of the distribution of the multiple correlation coefficient.

Non-coverage: Failure to include units or sections of the survey population in the operational sampling frame.

Non-determination: Coefficient of (k^2) Refers to a measure of alienation, which is the square of the coefficient of alienation, and related to the product moment correlation coefficient (ρ).

$$k^2=1-\rho^2$$

The coefficient measures the proximity of relationship between two varieties to zero- correlation.

Non-linear correlation: This term has been meant to relate to the correlation between varieties where the regression has been not linear. It is thus a misnomer; correlation, being a pure Dumber, cannot be non-linear.

Non-linear regression: The expression for the regression line may be non-linear in the coefficients; or in the variable, e.g., a polynomial relationship.

Non-normal population: Refers to a population for which the frequency distribution has been not the Normal distribution. The term does not imply abnormal in the sense of unusual.

Non-null hypothesis: In general it refers to a hypothesis alternative to the one under test; the null hypothesis. In some context, however, it has been given the meaning of a hypothesis under test where the effect has been not equal to zero.

Non-orthogonal data: Data have been said to be non-orthogonal if they give rise to estimates of various effects which are not independent of one another or, perhaps or other features such as block differences which are a nuisance in the analysis.

Non-parametric statistics: That branch of statistics comprising techniques of statistical analysis that do not entail assumptions about the exact form of the distribution of the population. The use of nonparametric statistical tests does not require that the sample being analysed be from a population with a normal distribution or with any other specified distribution that is, the exact shape of the distribution of the population need not be known.

Non-parametric tests of significance or distribution-free tests of significance: Tests of significance which make no assumptions concerning the parent distribution. All tests involving the ranks of data have been non-parametric.

Non-parametric tolerance limits: Tolerance limits which are not dependent on the parameters of the parent population from which a sample is drawn.

Non random sample: A sample which is selected by a non-random method. For example, a scheme whereby units have been selected purposively would provide a non-random sample. Again, a sample obtained by taking members at fixed intervals on a list has a non-random sample unless the list get arranged in a random order.

Non-response: The failure to get the required information from members of the selected sample for any reason. In order to measure the rate of non-response it becomes necessary to keep careful and complete records of the selection of individuals and the outcome of the fieldwork. The non-response rate may be defined as the proportion of designated individuals for whom the survey information was not obtained.

Non-sampling errors: Refer to errors in estimates which have been not due to the fact that the Information has been based on a sample from the population.

Non-singular distribution: Refers to a multivariate distribution in, say, p varieties which cannot, by a linear transformation of the varieties, be converted into a distribution with fewer than *p* varieties.

Normal curve, normal distribution: Refers to the most commonly used type of continuous frequency distribution.

Normal curve of error: A symmetrical frequency-distribution curve; variously called the normal probability curve, the normal frequency-distribution curve and, in its context, simply the normal curve.

Normal deviate: Refers to the value of a deviate of the Normal distribution.

Normal distribution: Refers to a continuous distribution of a random variable with its mean, median and mode equal. Symbol: N (μ, σ^2).

The probability density function $f(x)$ of the normal distribution, with mean μ and standard deviation σ has been given by

$$\phi(x) = \left(\frac{1}{\sqrt{2\pi\sigma}}\right) \exp\left\{-\frac{(x-\mu)^2}{2\sigma^2}\right\}$$

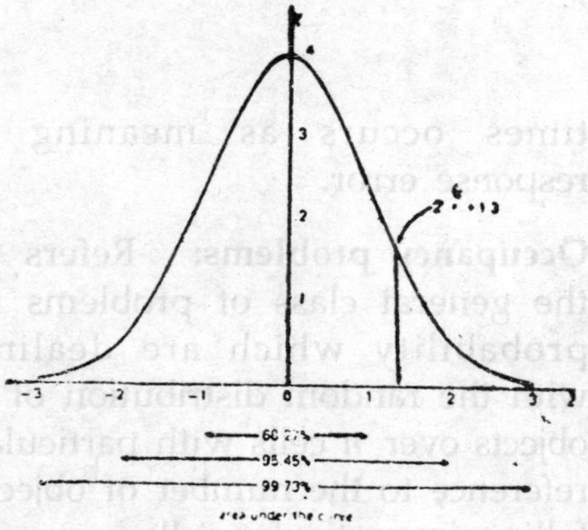

The distribution has been denoted by N(μ, σ²). When it has been given in standardized -form, with mean 0 and standard deviation 1, the distribution is denoted by N(0, 1), and has probability density function given by:

$$\phi(x) = \frac{1}{\sqrt{2\pi}} e^{-\frac{1}{2}x2}$$

The graph of the probability density function of the normal distribution has been a continuous bell-shaped curve, symmetrical about the mean, called the normal curve or error curve.

The-normal distribution has been very important in statistics for two reasons:

(a) It is the distribution of many naturally-occurring variables, such as the heights of adult men in a town, the masses of carrots in a field, etc.

(b) The distribution of the means of samples drawn from most patent populations is normal or approximately so, when the samples are sufficiently large.

Normal equations: A set (if simultaneous equations derived in the application of the least squares method, for example in multiple regression analysis. They are used to estimate the parameters of the model.

Normalisation of frequency function: Refers to the transformation of a variety so that its frequency function becomes normal, or approximately so.

Not applicable/does not, apply: A classification which is used to distinguish the situation where a question is not applicable to a particular respondent from a refusal to answer the question

Nyquist interval: For time series having a series of harmonic components with frequencies in a limited range, the nyquist interval has been such that the highest admissible frequency has a period of two intervals.

O

Oblique factor: In educational or psychological testing it refers to a factor which gets correlated with one, or more, other factors has been said to be oblique. In one geometrical representation on the situation the vectors which represent them have been no longer orthogonal one to another.

Observable variable: Refers to a mathematical or stochastic variable, the values of which could be directly observed, as distinct from unobservable variables which enter into structural equations but have not been directly observable.

Observation: (1) The observed value of a variate for an element is termed as an observation; and (2) observation is the process of getting information directly, for example, by observing behaviour —as distinct from obtaining information from reports and records by asking people about their behaviour, for example.

Observational error: Refers to an error of observation but sometimes occurs as meaning a response error.

Occupancy problems: Refers to the general class of problems in probability which are dealing with the random distribution of r objects over n cells with particular reference to the number of objects falling in particular cells.

Octiles: Refer to the set of seven variate values which would divide the total frequency into eight equal parts.

Odds: Refer to the ratio of the probability of an event occurring to that of its not occurring.

Odds ratio: Refers to an alternative term for relative risk.

A technique for specifying the nature of an office/plant for maximum system performance.

Ogive: Refers to a continuous form of the cumulative frequency curve which consider a very large number of observations. If the culmunative frequencies have been plotted against the ordinate axis and the variate values have been plotted against the abscissa, any unimodal culmunative frequency distribution will have an S-shaped appearance, taking the form illustrated.

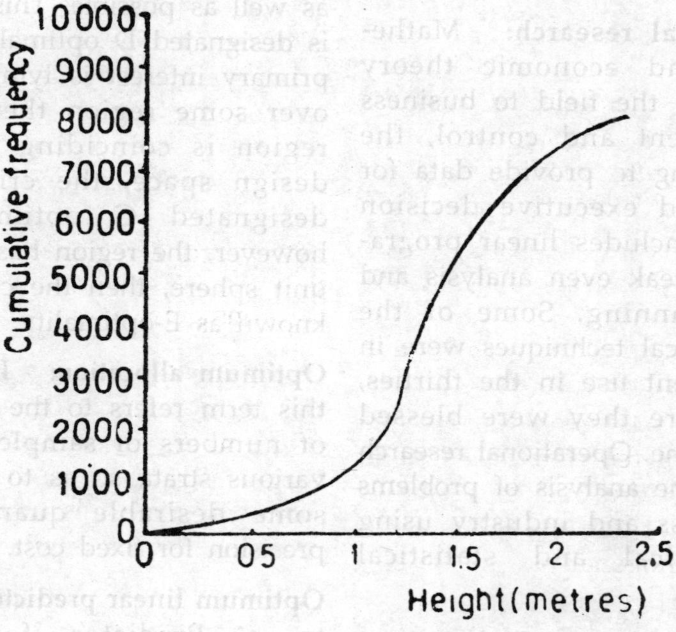

An ogive which shows the measurement of people under and including a given height.

One-sample runs test: A nonparametric method for determining the randomness with which sampled items have been selected.

Open-ended class A class that allows either the upper or lower end of a quantitative classification scheme to be limitless.

Open sequential scheme: Refers to a sequential sampling scheme which does not but an ultimate limit to the size of the sample.

Open ended classes: When in a frequency distribution, the initial class interval has been indeterminate at its beginning and for the final class interval has been indeterminate at its end, the distribution has been said to possess 'open ended' classes.

Open question, open-ended question: In an open question it has been the respondent who decides the form, length and content of the answer. The interviewer must record as much of the answer as possible.

Operating characteristic curve: In quality control, this curve refers to the graph of the probability of accepting a batch from which a sample has been drawn for inspection, against the true proportion of defectives for the batch.

Operational research: Mathematics and economic theory applied to the field to business management and control, the object being to provide data for policy and executive decision making. Includes linear programming, break even analysis and profit planning. Some of the mathematical techniques were in management use in the thirties, long before they were blessed with a name. Operational research involves the analysis of problems in business and industry using mathematical and statistical techniques.

Opinion survey: Refers to a sample survey which aims at ascertaining or elucidating opinions possessed by the members of given human population with regard to certain topics.

Optimal asymptotic test: An alternative name assigned to a locally asymptotically most powerful test

Optimality of design: In the design of an experiment of regression using the linear model, it has been possible to adopt certain desirable criteria which have been optimal in same sense. One natural criterion for parameter (s), a given confidence coefficient and residual sum of squares arising from a particular design, has been to make the content of the confidence ellipsoid as well as possible. This criterion is designated D optimality. If the primary interest is lying in $\xi(y)$ over some region then, if this region is coinciding with the design space, the criterion is designated G optimality. If however, the region has been the unit sphere, then the criterion is known as E-optimality.

Optimum allocation: In general, this term refers to the allocation of numbers of sample units to various strata so as to maximise some desirable quantity like precision for fixed cost.

Optimum linear predictor: Refers to a Predictor for future observations of a stochastic process which gets constrained to be a linear combination of past observations and which minimises the mean square error of prediction.

Optimum statistic: Refers to an expression which has been usually synonymous with best estimator. If it refers to a statistic used for testing hypotheses it is usually termed as an optimum test statistic.

Optimum stratification: Refers to a system of stratification in sampling which optimises some given criteria.

Optimum test: Refers to a test

which can be shown to have a certain desirable characteristic or group of characteristics to a greater degree than any other test of the some class.

Optimum value: Refers to that value which has the effect of maximizing another value, e.g., the optimum price and the EOQ, both of which have the effect of maximizing-profit.

Ordered categorisation Refers to a categorisation in which, although the variable has been not expressible in terms of an underlying measurable variable, it may nevertheless be arranged in order.

Ordered series: Refers to a set of variate values which have a natural sequence in time or space.

Ordinal number: Refers to a number which denotes order, quality or degree in a group, such as first, second, third.

Ordinal scale: Refers to a scale with classification and rank. However, equal differences in position on the scale do not necessarily represent equal differences in the variable being used for classification.

Ordinate: Refers to the vertical axis of a graph, the corresponding axis (or base line) being termed as the abscissa, or sometimes as the coordinate axis.

Overcoverage: Including elements for consideration which do not belong to the defined survey population.

Overhead: Of the total annual cost, that part which does not vary with the rate of output, but exists in time. Variously called fixed costs, oncosts, burden, time costs and period costs. The literature is often somewhat vague about the definition.

Overlap design: Refers to an alternative name which was given for serial designs in the context of routine quality control and experiments.

P

Paired sample's: 1 A sample in which the same attribute, or variable, of each number of the sample has been measured under different circumstances. 2. Two samples in which the members of each have been clearly paired.

Paired-sample test: Refers to a test on the differences of the random variable measured in a paired sample, usually of the Null Hypothesis that the mean value of the difference is zero.

Paired selection sample design: Refers to a sample design in which two units have been selected from each stratum. It is a form of replicated sampling.

Paired selections: The name assigned to two units selected randomly from within a stratum. Paired selections are forming the basis of a paired selection sample design.

Palm's theorem: It is a limit theorem which explains why for so many practical situations in which a stationary process has been well approximated by a renewal process, the renewal times have been negative exponentially distributed.

Parabola: An archlike curve described on the surface of a cone by the intersection, at a place below its apex, of a plane parallel to the opposite side. Now generally used of any archlike curve, such as that described by the Right through the air of zip, arrow or bullet, and that described by a quadratic equation whose coefficient of χ^2 has a negative sign. The graph of a frequency distribution is not called a parabola.

Parallel line assay Refers to an important method for the bioassay of a test preparation against a standard. If the expected response gets linearly related to, the logarithm of dose the regressions of response on log dose for the two preparations would often be parallel and the distance between them would estimate the logarithm of the relative potency.

Parameter of location (or scale):

Refers to a parameter of a frequency function which could be identified with some measure of location (or scale).

Parameter point: If a class of frequency functions is dependent on certain parameters ; e.g., if the unvariate function of x is $f(x, \theta_2, \theta_2, ...\theta_k)$, the domain of variations of the θ's is termed as the parameter space and any particular set of θ's determines a point in that space.

Parametric hypothesis: A hypothesis which concerns the parameter (a) of a distribution.

Parametric programming: Refers to a development of linear programming in which the parameters of the objective function, formed by the constraints, have been allowed to vary in a determinate fashion.

Parametric statistics: That brand of statistics which is comprising techniques of statistical analysis and require that the data being analysed be drawn from a population distribution with a specified form

Parent population: Refers to 'he population From which a sample has been drawn, any member of which could have been selected sometimes this is just referred to a the population.

Pareto curve/distribution: Refers to an empirical relationship which describes the number of persons Y whose income is x. It was first advanced by Pareto in the form

$$y = Ax^{-(1+a)},\ 0 \leq x \leq \infty$$

Pareto index: The coefficient a in the expression curve is generally termed as the Pareto Index. It affords evidence of the concentration of incomes, or, more generally, of the concentration of variate values in distributions of the Pareto type.

Pareto technique: A method of testing data to discover whether the Pareto principle holds good and of using the data into principle states that there is frequently an inverse relation between the percentage of items in each of a set of subclasses and the importance of the subclasses.

Pareto-type distribution: Refers to a loosely used expression which is used to denote any distribution shaped similarly to the, of Pareto. It is probably better avoided unless the distribution has been actually Paretian apart from origin and scale.

Parity equality: The term 'parity' is More commonly used in expressing equality of two sets of terms in an equation.

Part assembly sequence: When a part method is used in training, there are various ways of assembling the parts so as to ensure that newly learnt parts are integrated with parts learnt previously.

Partial association: Refers to the measure of association in subpopulations of qualitative characteristics analogous to tile partial correlations between quantitative variates. For example, if A, B and C have been three attributes the partial association of A and B with respect to C can be the association of A and B in the sub-population of members bearing the attribute C. Partial Confounding. If, in a factorial experiment with several replicates, there have been interactions which have been confounded in some replicates but not in others these interactions have been said to be partially confounded.

Partial correlation: Refers to the correlation between two vitiates, controlling for one or more other varieties.

Partial regression: In regression analysis, this form, refers to the coefficient of an 'independent' variable in the complete regression equation, that is to say, the regression equation which is involving all the Variables Under consideration, It is termed as partial, not a wary fortunate expression, because it differs from the coefficient which would be obtained if certain, perhaps all, other independent, variables get ignored and a simple regression of the dependent variate calculated in the one independent variable.

Partial serial correlation: Correlation within a series after permitting the influence of observations within the limits of the lag; it has been analogous to the ordinary concept of partial correlation, and is also known by the alternative name partial auto-correlation.

Partition of chi-squared (χ^2)**:** In certain circumstances the sum of squares of standardised Normal varieties about their mean, which gets distributed as χ^2, could be divided into two or more parts each of which is also distributed as χ^2 independently of the others This is termed as a partition of χ^2.

Pascal's distribution: It is given by:

$$p(x) = {}^{x-1}C_{x-n}\, p_n q^{x-n}$$

for $x>n$ where $0<p<1$ *and* $q=1—p$. This is the probability distribution for the number of bernoulli trials x needed to achieve n sucesses. Each trial has probability p of success, q of failure.

The mean and variance of this distribution have been given by:

$$\text{Mean} = \frac{nq}{p}$$

$$\text{Variance} = \frac{nq}{p^2}$$

When $n = 1$, Pascal's distribution becomes the geometric distribution.

Pascal triangle: A triangle of numbers named after Pascal to whom it gets attributed, showing the coefficients, 0 binomial expression raised to the power of k (k = 1,...n, where n has been the number of rows of the triangle).

Path coefficient: The coefficients of the variables in path models are termed as path coefficients, they represents the direct effect of one variable on another in the model. Conventionally the variables in the model have been assumed to be standardized.

Path diagram: Refers to the diagrammatic representation of a path model. The relationships between the variables are shown on the diagrams by lines, two kinds of relationships can be represented directed arrows indicating a causal relationship in the direction in which the arrow is pointing; and curved lines indicating simply an association between two variables.

Path model: Refers to a causal model in which the relationships between the variables have been represented by magnitudes of effects along paths connecting the variables in the model. The model could be represented either as a path diagram or as a system of equations.

Pearl-read curve: Refers to another name for the logistic curve, a general form of growth curve.

Pearson coefficient of correlation: The product moment coefficient of correlation has been sometimes referred to as the Pearson coefficient of correlation due to K. Pearson's part in introducing it into general use.

Pearson curve: Refers to a distribution from the family of frequency distributions which was developed by K. Pearson. The basic equation of the family is:

$$\frac{df}{dx} = \frac{(x-a)f}{b_0 + b_1x + b^2x_2}$$

where f denotes the frequency function. The constants of this equation could be expressed in terms of the first four moments, if these exist. The explict solutions have been classified into types according to the nature of the roots of the equation $b_o+b_ix+b^2x_2=0$.

Peek group: Refers to a social group which is composed of individuals of approximately the same age or status.

Pentad criterion: In factor analysis, this term refers to an extension of the tetrad criterion which was developed by Kelley and Holzinger and based on sets of five correlations from the correlation matrix.

Percentage: Refers to the proportion or rate per hundred parts. Symbol: %,. To write a number as a percentage.

Percentage bar diagram: It is a method of showing a population which has been divided into several sets.

Percentage diagram: Refers to a diagram which shows a simple analysis of statistical data in terms of percentages. The actual form of the diagram could vary; examples have been the bar chart and the pie chart.

Percentage distribution. Refers to a frequency distribution having -the total frequency equated to one hundred and the individual class frequencies expressed in proportion to that figure.

Percentage laws: The percentage rate per unit to which a dependent variable tends to rise or fall when an independent variable is doubled. The eighty per cent law said to apply to learning is well known: it states that when the cumulative output of a given product doubles, the labour time per unit of output falls to 80%. The laws apply only to relationships that are logarithmic.

Percentiles: The kth percentile is defined as the value of the data such that k per cent of the observations are less than that value and 100-*k* per cent are greater than that value. For example, the 50th percentile is called the median.

Perfect-competition: A market structure is perfectly competitive if the following conditions hold; there are a large number of firms each with an insubstantial share of the market. These firms produce a homogeneous product using identical production processes and possess perfect information. It is also the case that there is free entry to the industry, that is new firms can and will enter the industry if they observe that greater than normal profits are being earned. The effect of this free entry is to push the demand curve facing each firm downwards until each firm earns only normal profits, at which point there is no further incentive for new entrants to come into the industry.

Performance and cost indices: A performance index is a ratio of two quantities, one of which represents the output or part of the output and the other the input or part of the input of a system.

A cost index is a special type of performance index in which the input is a cost and the output may be expressed in monetary units or physical units.

Performance rating. Performance in behaviour when related to a standard. Performance rating is the rating of the performance of a person or group.

Periodicity: Refers to the

characteristic of happening at regular intervals in time. The term not only relates to cyclical fluctuations and seasonal fluctuations, but sometimes to any wavelike oscillation which has identifiable maxima and minima for each of a series of cycles.

Periodic process: If any realisation of a stationary stochastic process yields a series which has been strictly periodic then the process has been a periodic process.

Period of a state: The period of a return state k of a Markov chain has been the greatest common divisor Ak of the set of in tegty n for which f_{kk} $(n) > 0$, where f_{kk} (n) has been the probabilers that the (first) return time of state k is n.

If λ_k has been greater than unity, state k has been termed as 'periodic' ; if $\lambda_k = 1$ the state has been called aperiodic.

Permissible estimator: Refers to an estimator which yields permissible estimators for all possible samples; permissible meaning that the estimates lie in the known range of the parameter.

Permutation: Refers to an order in which a given set of objects can be arranged. Thus, for example, the set ABCD may be ordered in the following permutations

ABCD BACD CBDA DABC
ABDC BADC CBAD DACB
ACBD BCDA CABD DBCA
ACDB BCAD CADB DBAC
ADBC BDAC CDBA DCAB
ADCB BDCA CDAB DCBA

Note that there have been twenty-four arrangements, or permutations, of the set ABCD. It can be expressed as $4! = 4 \times 3 \times 2 \times 1 = 24$.

The number of permutations (of r elements) that can be obtained from a set of objects of size n is

$$\frac{n!}{r!(n-r)!} Xr!$$

that is, the number of combinations of size r multiplied by the number by the number of permutations of each r elements. This simplifies to:

$$\frac{n!}{(n-r)!}$$

$$\frac{10!}{6! \times 4!} \; 4! = 210 \times 24 = 5040$$

or alternatively

$$\frac{10!}{6!} = 10 \times 9 \times 8 \times 7 = 5040$$

The term permutation is often misused by those who fill in the football pools #nd use it to mean combination.

Permutation tests: Refer to a class of distribution free tests which are based upon the fact that any ordering of a random sample of n items is having the same probability $1/n$!.

Persistent state: Refers to a recurrent state for which the average return time has been finite.

Peters' method: Refers to-a method of estimating the standard deviation of a distribution which has been approximately Normal by multiplying the mean deviation by 1.253, this being the ratio $\sqrt{(½\pi)}$ appropriate to the Normal distribution.

Phase: The interval between the turning points of a series which is ordered in time or space is called a phase. The distribution of phase lengths provides one test of random order.

Phase spectrum: The sample phase spectrum reveals whether the frequency components in one of two series lead or lag the components at the same frequency in the other series ; it has been a description of the covariance between the two series.

Pictogram A method of pictorial representation of quantities or magnitudes in a graph, bar-chart, band-chart or cartogrարn by using a number of miniature drawings of the subject matter, or of related symbols, e.g., $ signs for pounds sterling, barrels for crude oil production, and so on.

Pie Diagram: A more picturesque term for the circular diagram or chart.

Piecewise regression: Refers to an alternative name for segmented regression.

Pivot: Refers to the essential element in the inversion process which is necessary for the solution of a linear program.

Planned economy: An economy where crucial economic processes are determined to a large extent not by market forces, but by an economic planning body which implements society's major economic goals. In such a system the greater part of production activity is determined by obligatory input and output targets generated by vertical signals from an administrative hierarchical body. The advantages of such a system over the pure market economy include the reduction of uncoordinated economic activity which may lead to an inefficient utilization of available resources and the production of undesirable commodities.

Plot: In experimental design this term is used to refer to the basic unit of the experimental material. Although it derives from the physical unit of a plot of land in an agricultural trial, its interpretation has been very much more general according to the subject matter of the particular design.

Point binomial: It is an alternative name for the Bernoulli, or binomial, distribution. The word 'point', which-is not necessary, arises from the discrete character of the variate.

Point biserial correlation: Refers to a modification of the biserial correlation to the case where one variate, instead of being based on a dichotomy of *n* underlying continuous variate, has been discontinuous and two valued.

Point bivariate distribution: An alternative name for a bivariate distribution of two discrete variates

Point density: Refers to the relative frequency, or probability mass, which may get located at the different point values of a discontinuous variate. The term has been better avoided favour of frequency or probability mass, the word 'density' being reserved for continuous probabilities.

Point estimation: Refers to estimation of a parameter of a parent population as a single value. This contrasts with interval estimation, where the parameter is estimated to lie within an interval.

Point-of-time Statistics: Statistics that relate to a point of time as distinct from rate-of-flow statistics. Point-of-time statistics include stock figures, the book values of fixed assets price, marginal cost marginal revenue and unit cost; rate- of-flow statistics cover figures that are expressed per unit of time (day, week, year, etc.), such as operating costs, sales, annual costs and fixed costs.

Poisson binomial distribution: Refers to a discrete distribution of the number of successes in 'n' independent trials having probability *pj* of success in the jth trial. The *binomial* distribution has been a special case where the parameters are $n, p_1, p_2, p_3, \dots p_n$ and $p_1=p_2=p_3=\dots=p_n=p$.

Poisson distribution: Refers to a special type of probability distribution which can be applied to a large number of physical and natural phenomena. Suppose there have been a large number of occasions during a particular period of time in which some event can occur, and on any one occasion there has been a small but identical probability that it will occur, but the average rate of occurrence for identical periods is the same, then the Poisson distribution would provide the probabilities that during the period of time the event will occur a, specified number of times.

Poisson-normal distribution: Refers to a distribution-which resulted from the expected value of a Poisson distribution, itself taking a truncated Normal distribution. It is also sometimes termed as the Gauss-Poisson distribution.

Poisson-Pascal distribution A name which is sometimes assigned to the distribution obtained when a negative binomial distribution and a Poisson distribution have been convoluted.

Poisson process: Refers to a stochastic process which may be considered as a particular case of a birth process in which the parameter λ_n has been a constant λ.

Politz-simmons method: Refers to a technique which tries to eliminate the need for call- backs in social surveys by using additional information available for those contracted in the' first cell.

Polya process: Refers to a particular case of a birth process in which the parameter A. has been given as follows

$$\lambda_n(t)=(1+an)/(1+at),$$

a being a constant.

Polychoric correlation: Refers to an extension of tetrachoric correlation to the case of an m by n table where it may be regarded that the two underlying variates have been jointly normally distributed.

Polyspectra: The spectrum of a time series could be considered as the Fourier transform of its autocorrelations, which are dependent on the product of terms $u(t)u(t+k)$. The transform of the product of more than two terms has been termed as a polyspectrum, e.g., for three terms we are having the bispectrum.

Polytomic table: Refers to a contingency table having more than two categories in the row and column classifications.

Polytomous: Divided into many parts, or categories; as distinguished, for example, from dichotomous, which means divided into two parts.

Population: The word population, when used by a statistician, may refer to any specified collection of objects, people, organizations, etc. Some example,, of a population are (1) individuals in a country, town or constituency; (2) animals on a farm; (3) commercial firms in a city; (4) record cards in a filing system; (5) children in a school; or, indeed, any other predefined set of elements. The population is the parent group.

The term 'population' has displaced the older term 'universe', which itself derived from the 'universe of discourse' of logic. It is practically synonymous with 'aggregate' and does not necessarily refer to a collection of living organisms.

Population closed: A population 'into which there is no immigration and from which there is no emigration. Population change, therefore, is determined entirely by the relationship between the birth rate and the death rate.

Population, finite: In statistics, an existing (or previously existing) population with a fixed number of cases. A finite population may, but does not necessarily, consist of persons. For example all the automobile accidents in United States during a given five-year period would be a finite population,

Population, infinite: In statistics a theoretical population with an unlimited number of cases. An example of an infinite population would be all the possible tosses of a coin. Population

Population profile: A type of bar chart which is used to display the population age structure of a country (or city, region, etc.). The population figures have been divided into age groups, and into male and female. The scale for the bars may be for the actual numbers or for their percentages.

Positive binomial distribution: Refers to the distribution which is formed from the binomial distribution by omitting the value 0. The name is also sometimes assigned to the usual form of the binomial distribution so as to distinguish it from the negative binomial distribution.

Posterior probability: The term means an estimate of probability after a given event. In contrast, prior probability is the estimate of probability before the event. The terms are relative.

Power: The term has two different meanings

1. An exponent of logarithm indicating the number of times which a base number must be multiplied by itself to obtain a given value, for example the value 128 is 2^7 (i.e., 2 multiplied by itself 7 times), thus the base number 2 is raised to the power of 7 ; and
2. In statistical testing the power of a test is 1-β, where β is the probability of type II error; that is, its power is the probability that the alternative hypothesis will be rejected when it is false.

Power curve: A graph of the values of the power of a test for each value of μ, or other population parameter, for which the alternative hypothesis is true.

Power efficiency: Alternative name for relative efficiency of a test.

Power function: The power of a test of the null hypothesis, expressed as a function of a parameter 0 which expresses The class of alternatives to it.

Power of the hypothesis test: The probability of rejecting the null hypothesis when it is false, i.e., a measure of how well tile hypothesis test is working.

Power statistical: The probability that a given statistical test will reveal a statistically significant relationship when such a relationship really does exist.

Power sum: Refers to the sum of a series of observations on a variate each of which is raised to the same power. Such qualities are found most frequently in the calculation of moments or similar symmetric functions of the observations.

PPS: A widely used abbreviation for probability proportional to size.

Pre-coded questions: Questions to which the response has been directly coded into one of a limited number of categories.

Predetermined variable: In the statistical analysis of models, especially of the economic kind, it is possible to classify the variables as endogenous or exogenous according to whether they represent an integral part of the system or influences impinging on it from without.

Prediction: The term usually implies the estimation of a future value of a variate either (1) by the projection of a trend regression line in time series analysis or (2) by the use of a probabilistic model or (3) both.

Prediction error of: Refers to the difference between a prediction and the true value of the identity predicted.

Prediction interval: Refers to the interval between the upper and lower limits attached to a predicted value to show, on a probability basis, its range of error.

Predictive validity: If a scale has been constructed as a predictor of behaviour or as an indicator of future performance its success (or validity) could be measured by the degree to which it succeeds in this prediction or indication. This is termed as predictive validity.

Preference table: The (21) paired comparisons of the objects, which ties are not permitted, may be displayed by means of 0, 1 variables in a two way table termed as a preference table.

Price compensation index. Refers to an index number-of consumers prices which is constructed as a chain index on the basis of a consumer income and varies so as to maintain a constant standard of living. The Laspeyres- Konyus index has been of this type.

Price discrimination: The charging of different prices to different customers; usually results from differences in size of order and delivery, and often takes the form of a discount for large orders.

Price index: Refers to an index number which purports to combine several series of price data into a single series expressing an average level of prices; e.g., of retail prices or of prices of manufactured products.

Price index: The ratio of the average of prices in one period or place (referred to as the given period or place) to the average in another period or place (referred to as the base period).

Primary data: Data which are obtained direct from source and features have been : (1) that they are original, having been obtained by specific problem in hand. Their essential features have been obtained by specific investigation rather than from a published, classified or structured source; and (2) that they are obtained for a specific purpose relevant to the enquiry which has been carried out. The principal means of obtaining primary data have been by self-administered questionnaires, interviews, observation and experiment.

Principle of equipartition: Refers to the division of the range of a frequency function into a number of parts such that the frequencies corresponding to each part have been equal, This method has been sometimes used in tests of goodness of fit I it has been also used in sample surveys for the construction of strata.

Principle of the irrelevant question: Refers to a technique for getting information about a population which might be damaging or incriminating to individuals.

Prior probability. (The prior probability of an evert A, with respect to event B). Refer to the probability that event A occurs if it is known if event B has occurred or not. This is thus p (A). The posterior probability, for the same pair of events, has been the probability that event A occurs if it has been known that event B has occurred. This is written p(A I B).

Priority queueing: Refers to a queueing system where the actual order of service for the arrivals has been determined by some scheme of relative priority, i.e., not a first come first served discipline.

Probability: This term must be understood from two standpoints, its mathematical foundation and the interpretation d probability measurements.

The mathematical foundation of probability in set theory had been laid by Kolmogorov and others. Probability has been measured in the scale from 0 (impossible) to I (certainty). If two events have been equally likely and one or other of them is certain to happen (such as the outcomes 'head' or 'tail' on tossing an unbiased coin), the probability of each of outcome has been 0.5. Similarly, if a die has been cast, given that it is also unbiased, since it consists of six sides which are equi-probable and it-is certain that one side will land uppermost, the probability measurement in respect of each of the outcomes is 1/6.

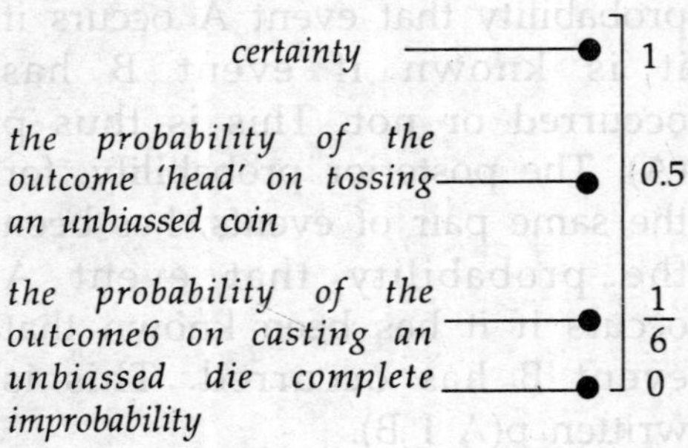

Probability density function: Refers to an alternative term for the frequency function when the distribution concerned is considered as one of probability. It is a mathematical expression showing the relative frequency (or probable density) of *x* as a function of *x, i.e., f(x)*.

Probability distribution: Refers to the distribution of the probabilities of occurrence of the values of a variate as distinct from a frequency distribution, that provides the frequencies of the values of the variate. Probability Element. The probability which is associated with a small interval of a continuous variate and written in some such form as f (x)dx, or in general;

$$f\,(x_1, x_2,...., d_n)\,(dx_1, dx_2,...dx_n.$$

Probability-generating function: Refers to an expression which, when expanded, has terms whose coefficients have been the probabilities of !he different numbers of occurrences of an event.

The probability-generating function for the binomial distribution with probability p of success, $q=1-p$ of failure, has been given for n trials by:

$$(q+pt)^n$$

where t denotes a success.

Probability integral: Refers to an alternative name for the distribution function or the cumulative probability function continuous variates. For example, the probability integral a continuous variate x has been a function $f(x)$ having the property that

$$F(a) = Pr\{x \leq a\} = \int_{-\infty}^{a} f(x)\,dx$$

where $f(x)$ has been the probability (frequency) function.

Probability integral transformation: If x has been a continuous variate having frequency function $f(x)$ and distribution function f(x), a transformation to a new variate y given by

$$y = \int_{-\infty}^{a} f(x)\ dx = F(x)$$

has been termed as the probability integral transformation, y has been uniformly or rectangularly distributed in the range $0 \leq y \leq 1$.

Probability limits: Refer to upper and lower limits which are assigned to an estimated value for the purpose of indicating the range within which the true value has been supposed to lie according to some statement of a probabilistic character.

Probability surface: Refers to a bivariate frequency surface; that is to say, the three dimensional representation of a bivariate frequency (probability) distribution with the frequency (probability density) along one axis and the variates along the other two axes.

Probability tree: Refers to a graphical representation which is showing the Possible outcomes of a series of experiments and their respective probabilities.

Probable error: A statistical measure of the error that results from the differences between a sample and the total Population from which the sample was drawn. The probable error of the mean (P.E.) is the interval in a frequency distribution that centers about the arithmetic mean and includes 50% of all the cases in the distribution, extending from that point which includes half the cases below the mean' (25% of all the cases) to that point which includes half the cases above the mean (25% of all the cases). In a normal frequency distribution, this is equal to the quartile deviation.

Probit analysis: Refers to the analysis of quantal response data which is using the probit transformation.

Procedural bias: Refers to a bias in a sampling inquiry attributable to the procedure followed in obtaining the information, as distinct from bias due to the use of inferior methods of estimation on the data when obtained.

Process average: In quality control, this term refers to the proportion of defective items produced by a machine or manufacturing process when it is running normally.

Process average proportion defective: The term is ;used in quality control to indicate the average proportion of defectives in a batch or sample, and thus the probability that any individual item produced in a controlled process is defective.

Processing error: Refers to a type of error which can occur in the processing of statistical data. In survey data, for example, processing errors may be including errors of transcription errors of coding, errors of punching on to cards and errors of arithmetic in tabulation,

Producler's risk: The risk that a particular batch of satisfactory quality will be rejected by a sampling plan. The risk will depend on the level of quality required and the predetermined criteria of the sampling plan.

Productivity: Unqualified, the output per man hour, man month or man year. Where it refers to a factor of production other than labour, it is usually qualified. The productivity of capital is some what Ambiguous, and where it refers to the output of a particular type of machine or vehicle, it is probably best expressed in terms of labour productivity.

Product mix: The body of ratios of the various products made or sold by a company to the total in terms of value. Optimum product mix is that body of ratios which maximizes the company's profit. It is a technique of linear programming but, as usually expounded, it makes a number of unrealistic assumptions, e.g., that 'the company's assets can be switched from producing one type of article to producing any other of the company's range and that there is no limit to the demand for any product at a fixed price.

Profile analysis: Refers to a method of clustering the attitude questions in a survey. In common with factor analysis and linkage analysis, it begins from the correlations between the pairs of attitude questions, arranged in the usual correlation matrix. The Profile of an attitude question has been its correlation with the other attitude questions, i.e., its row in the correlation matrix.

Profit planning: Any planning designed to increase or at least maintain the company's profit could be described as profit planning. However, in the literature is has come to mean an extension of linear break-ever analysis.

Projection: This term finds use in two connected senses. (1) In relation to a time series it implies a future value which is calculated according to predetermined changes in the assumptions of the environment. (2) more recently it is used in probability theory to denote the conditional expectation of a variate.

Propensity: A quantification of a given trend or tendency in terms of proportion, probability or relative frequency. For example, in economic statistics the propensity to save means the proportion of a given income which will be saved , while the propensity to consume means the proportion of a given income which is likely to be consumed.

Proportionate stratified sampling: Stratification represents a procedure by which the population has been divided up into subdivisions, known as strata, in each of which sampling is carried out independently. Proportionate stratified sampling involves selecting a number of units from each stratum which is proportional to the number of units in the stratum; in other words, we would use the same sampling fraction in each stratum.

Proximity analysis: A method which is used in numerical taxonomy for arranging items in a line or a plane, or in a space of higher dimensions so that like has been adjacent to like and far from unlike. The method has been based upon rank ordering of inter-point distances and inverse ordering of similarities.

Pseudo-factor: Refers to an artificial or dummy factor which is used in the design of factorial experiments, generally to render the number of factors a convenient one for the application of some specified balanced design.

Pseudorandom number: A number taken from a set of numbers which are not truly random, for example, because they are self-repeating or degenerate.

Pseudo-inverse: Refers to a mathematical device which tries to overcome the fact that a square matrix which is Singular does not have an inverse form.

Pseudo- spectrum: This term is used to denote the mathematical expectation of the estimates of spectral density obtained without regard to the stationarity of the process.

Psi-square statistic: Refers to a chi-squared type statistic which can be used as a lest of fit with serially correlated observations.

Public good: A commodity or service which if supplied to one person can be made available to others at no extra cost. A public good is thus said to exhibit nonrival consumption; one person's consumption of the good does not reduce its availability to anyone else. It may be contrasted with a private good where one person's consumption pre-cludes the consumption of the same unit by another person, The extreme, or 'polar', case of a 'pure' public good has been defined by Paul A. Samuelson as a good which is:

1. non-rival in consumption
2. has the characteristic of non-excludability—that is, if the good is provided the producer is unable to prevent anyone from consuming it.

This latter characteristic prevents private markets from functioning since a seller would be unable to ensure that only those individuals who paid for the good could obtain it. Since the good could be obtained without payment no-one would be willing to pay for it.

Purchasing power of money: The quantity of goods that a unit of money will purchase, usually expressed in index number to the community in general, it is the inverse of the retail-price index; to a manufacturer, it may be the inverse of the price index of the materials he purchases and the labour he employs.

Q

Q-technique: Refers to a method of analysis of similarities of relationships in which, given a matrix of *n* observations on *m* individuals, *n* rows and *m* columns, the correlations or other statistical measure have been sought between the *m* columns down the *n* rows i.e., between individuals, as distinct from the R-technique which is looking for relations between variables, namely between rows along the columns.

Quad: Refers to a square- shaped basic cell ; also, the area of such a cell.

Quadrant: One of the four quadrants of the graph. Each quadrant is identified by number or its compass position. The NE quadrant is no. 1 and is the one that is most used as it accommodates the positive values of X and Y

Sign of	*Quadrant*	
X	Y	
+	+	No. 1 : NE
−	+	No. 2: NW
−	−	No. 3: SW
+	−	No. 4 : SE

The horizontal axis is called the *x*-axis, and the vertical axis, the *y*-axis. Their point of intersection is called the origin, and is marked 'O' : it provides the position of the dot for Y = X = 0.

Quadrant dependence: If the probability of any quadrant $X \leq x$, $Y \leq y$ under the distribution $F(x, y)$ gets compared with its independence probability, then the pair (x, y) or its distribution as represented by Lehmann is positively quadrant dependent if

$$Pr(X \leq x, Y \leq y) \geq Pr(X \leq x)\, Pr(Y \leq y).$$

The negative dependence is holding with the inequality reversed.

Quadrat: Refers to an area, usually of vegetation, randomly selected for study. It is normally square in shape. The ideal size of a quadrat has been the smallest size that has same number of species as would be contained in a larger one. In a point quadrat sampling has been done at the Points of a square grid covering the quadrat.

Quadratic estimator Refers to an estimator which is based upon some quadratic function of sample values. For example, the standard deviation could be estimated from the square rout of the variance, a quadratic estimator, or from the mean deviation or the range, which have been linear estimators.

Quadratic mean: An infrequently used term which denotes any mean which gets derived by employing the squares of the values for which the mean has been produced.

Quadratic programming: Refers to a major development of linear programming in which some or all of the constraints and the objective function have been quadratic in the variables.

Quadratic response: Where the relationship between a quantitative response and the dose metameter has been not linear, but has been of the second degree, the response has been sometimes termed a quadratic response.

Quadrature spectrum: Refers to the form of spectrum which is able to measure the covariance, between the sine and cosine components, or out of phase components, of a sample time series. If the quadrature spectrum has been denoted by Q_{12} (f) then

$$Q_{12}\ (f) = -A_{12}\ (f) \sin F_{12}$$

where A_{12},(f) and F_{12} (f) have been the phase spectrum and cross-amplitude spectrum of the two series.

Qualitative data: Data that are not in numerical form. Qualitative data are data in the form of descriptive accounts of observations, inverviews, or written material. Data in qualitative form may be possible to quantify.

Quality control: Refers to the control of the relative quality of a manufactured product, usually by statistical sampling techniques. The reason for using statistical analysis has been that the product has been manufactured in such large numbers that complete inspection of all articles has been impossible. Samples from batches have been therefore analysed either; (1) with the object of discovering the number of defective items; or (2) with the object of comparing the mean and standard deviation of product size or component quantity with the desirable size or component quantity, given clearance limits.

Quantal response: The response of a subject to a stimulus has been said to be quantal when the only observable, of the only recorded, consequence of applying the stimulus has been the presence or absence of a certain reaction e.g.., death. A quantal response could be expressed as a two valued variate which is taking values 0 and 1.

Quantiles: Values as certain positions in a set of numbers a-ranged in order of magnitude. Common examples are deciles and percentiles.

Quantitative data: Refers to the data which is classified by some numerical value. Compare qualitative data. Thus, data on the number of hours of sleep recorded for certain hospital patients has been expressed numerically, and has been quantitative data.

Quantitative response: Refers to a reaction, by an experimental unit to a given stimulus, which could be measured on a variance scale. For example, the response could be measured in terms of weight, size or reaction time : in particular the survival time.

Quantity index: A measure of how much the number or quantity of a variable changes over time.

Quantity weights: When constructing a price index it has been normal to form a weighted average of the relative prices of the items covered by the index. These weights have been the quantities of the items purchased in some reference period, which could be either the current period for which the index has been calculated or some previous base period.

Quantum changes: Refers to the relative changes in quantities purchased between two periods which are used to construct a quantity index number. They play the same role in the quantity index as price changes play in a price index.

Quantum hypothesis: Refers to a hypothesis in which the possible values of a parameter have been discrete and hence increase by quantum jumps.

Quantum index: Refers to an index number which purports to show the changes in quantity, generally of goods or services produced, purchased or sold, independently of changes in prices or money values. One such index has been of the Laspeyres' type obtained by weighting quantities in the given and base period by prices in the base period.

Quartile: Refers to the value of a variable below which three or one quarter of a distribution lie. The median has been the second quartile.

Quartile deviation: Refers to a measure of dispersion which is based upon the distance between certain representative values of the variate. In this case the representative values have been

The tipper and lower quartiles and the quartile deviation is defined by.

Quartile, lower: In a frequency distribution, the first quartile or 25th percentile, below which lies one quarter of the cases.

Quartile, upper: In a frequency distribution, the third quartile or 75th percentile, below which lie three-quarters of the cases and above which lies one-quarter of these cases.

Quartile measure of skewness: Refers to an easily obtainable measure of skewness which uses the fact that in a skewed distribution the median does not lie exactly midway between the quartiles. The extant of skewness could be assessed by doubling the median and deducting this from the sum of the quartiles, $Q_3 + Q_1 = 2M$. To transform this measure into relative terms, it becomes necessary to obtain the coefficient of skewness, by dividing by the interquartile range. Thus

Quartile coefficient of skewness =

$$\frac{Q_3 + Q_1 - 2M}{(Q_3 - Q_1)}$$

It has been the most common quartile measure of skewness. The measure $Q_3+Q_1-2M/½(Q_3-Q_1)$ appears in some texts.

Quartile variation: This term is alternative to the standard deviation as a measure of variation. If the lower quartile been Q_1 and the upper quartile has been Q_3, the coefficient quartile variation, denoted V_q has been given by

$$100(Q_3 - Q_1)/(R_3 + Q_1)$$

Quasi-factorial design: Refers to an experimental design which a formal correspondence could be set up between treatments and their combinations and the combinations of treatments of a factorial set.

Quasi-Markov chain: Refers to a process over N+1 states identified by the integer 0 to N, for which the Markovian pro-perty holds between any two states i and j for which $1 \leq i, j \leq N$. Such a quasi-Markov chain has been said to be of order N. This chain arises in the theory of regenerative phenomena.

Quasi-maximum likelihood estimator: Refers to a set of least squares estimators for multiple equation models where the asymptotic distributional pro-perties of the estimators have been almost independent of the form of the distributions of errors in the equation, so that least squares solutions have been virtually equivalent to maximum likelihood solutions.

Quasimedian: Refers to a generalisation of the concept of a sample median to the arithmetic mean of any pair of symmetrically ranked order statistics. The usual definition of a sample median has been obtained by setting i to zero.

Quasi-normal equations: If the normal equations of least squares estimation are having the parameter estimates based upon instrumental variables rather than the ordinary model variables, then the equations would be said to be quasi- normal. 'Normal' in this sense is nothing to do with the 'Normal' (Gaussian) distribution.

Quasi-range: This term was given by Mosteller for the difference $x_s - x_r$ where $1 \leq r \leq s \leq n$, in the ascending order statistics of a sample of n observations. Common practice has been *to* take the range of $n - 2i$ observations, omitting the i largest and i smallest.

Quasi-random sampling: The employment of a method of sampling which is not truely random, but where the results will with confidence for practical purposes approximate time which would occur if a truely random sampling process were used.

Questionnaire: A document containing a list of questions, copies of which are sent to a number of people or organizations who are invited to provide the answers, or required to do so under the provisions of a statutory instrument, or under the membership rules of a trade association or similar body.

Questionnaire design: Refers to a formulated series of question, which are used for getting information on special points.

Queueing problem: The problem of queues, or congestion, arises in a variety of fields where there has been a service to be offered and accepted rather than a product to be made.

Queueing theory: This is concerned with the mathematical analysis of problems in which items requiring servicing or item providing service occasionally stand idle, and it ii necessary to specify either an arrival rate or service rate or both, in order to achieve the most satisfactory (least costly) solution.

Quota samples: A sample which is usually of human beings, in which each investigator gets instructed to collect information from an assigned number of individuals (the quota) but the individuals are left to his personal choice.

Quota sampling: This term, describes loosely a set of sampling methods which are used widely by market research organizations in interview surveys.

R

Radix: Refers to the number upon which a system of numbers has been based. For example, the radix of a system of natural logarithms has been (= 2.718)

Raising factor: This term is generally synonymous with one of the meanings of factor, that by which a quantity has to be multiplied to obtain another quantity. It is, however, some-times used specifically to mean the factor by which a sample value has to be raised to obtain a parameter estimate.

Random balance design: Refers to a factorial design which is intended to deal with the case where the number of factors exceeds the number of obser-vations. It has been con-structed by selecting treatment combin-ations at random from the possible set, subject to constraints to ensure balance.

Random bifurcation: Refers to the manner of growth of a tree or rooted graph when the free end that bifurcates on each occasion has been a random choice from those available.

Random component: If a magnitude is having a number of parts compounded in some way, e.g., by addition or multiplication, any such part as is a variate has been random component of the magnitude.

Random Error: An error in the deviation between the true value and the observed value or between the true value and its estimate.

Random impulse process: Refers to a stochastic process which is describing the linear motion of a particle subject to a series of small impulses which have been random.

Random numbers table of: A printed table listing numbers in random order that may be used in selecting a random sample. The arrangement of the directions.

Random sample: Refers to a sample which is selected by a method of random selection. The

American term 'probability sample' has been probably preferable as the term random may get be confuse with haphazard. The term random sample pie, although this usage is not satisfactory. Random samples have been distinguished from quota model sampling

Random series: Refers to a series the numbers of which may be considered as drawn at random from a fixed distribution.

Random standardized normal deviates: Refer to a set of random numbers which are converted into standard score (Z) equivalents.

Random start: In systematic sampling the first sample unit is selected at random from the first set of sampling units—the first sampling interval. The sample is then said to have random start.

Randomization: Refers to a process of ordering a set of object (or numbers using a process that ensures that every possible order (or arrangement of items) of the set has been equally likely.

Randomisation test: Refers to any test which is characterised by the test statistic being referred tinder the null hypothesis, to its distribution on permuting the observations in every equally likely way that could have occurred under that hypothesis.

Randomised model: Refers to a statistical model, generally an experiment design, where the treatment combinations have been assigned to experimental units by some random arrangement.

Range: Refers to the difference between the largest and the smallest values of a sample or set of variables.

Range chart: Refer to a chart which is used in statistical quality control on which the recorded quality criterion is the range of samples. This control chart has been used to maintain a check upon the variability of the quality of the particular product or processes.

Rank correlation: Refers to correlation of data by rank, rather than by numerical value. This can be measured using Kendall's or Spearman's rank correlation coefficient.

Random order statistics: Type of statistics which gets based only on the rank order of the sample observations, e.g., the rank correlation coefficients. Rank order statistics' have been distinguishable from 'order statistics', e.g., the median, the range which have made use of the metric values of the observations.

Ratio delay method: An alternative name for the snap reading method.

Ratio estimator: Refers to an estimator which takes the form of a ratio between two random variables. Such ratios are usually used in sample surveys. The value obtained by using a ratio estimator has been known as a ratio estimate.

Ratio scale: (1) Refers to a scale of measurement which has the properties of an interval scale together with an absolute zero or origin. (2) Refers to a graphical scale in which equal absolute variations correspond to equal proportional variations in the data.

Rational price fixing: The fixing of prices that maximize profits. The following formula is used to derive the optimum price, as the rational price is often called:

$$p_m = a\frac{e}{e-1}$$

where p_m is the optimum price, a the marginal cost, and e the price elasticity of demand. Its derivation is given under marginal revenue.

Raw data The data from a survey or experiment in their original form, i.e., before being modified or analysed in any way. Raw data are, however, often assumed to be edited and coded.

Raw score: The score as originally obtained in some psychological, educational or other test.

Recall errors: Memory Errors in answering a factual question the respondent usually relies to some degree on memory to provide an accurate answer.

Recall loss: In questions dealing with past events and therefore fail to report them. This has been recall loss.

Record, recording: Refers to substituting new codes for the original coding of the data.

Record check: A method which is used occasionally to check on the accuracy of the information collected in survey. There may be some events source (record) for some of the information obtained.

Records tests: Distribution free tests for trend in time which series based on the breaking of record values.

Rectified index number: Refers to an index number formula which has been obtained by taking the geometric mean of two other index numbers of opposite bias.

Rectifying inspection: Refers to the inspection of a product which aims at removing any defective units found and replacing them by effective units.

Rectilinear trend: An alternative name for a linear trend.

Recurrence game: Refers to a sequence of trials of an event which have been conducted as a game, in that a 'reward' has been received, or a 'fee' incurred, under certain recurrences.

'Recursive' system: The word 'recursive' is used by some workers to denote relations which have been recurrent presumably under the impression that the former is the correct adjective form derived from the verb to recur. A pursuit will avoid the word in this sense.

Reduced design Refers to a device for restricting the large size of block needed by a full design.

Reduced equations: Refers to a method of estimation in econometrics whereby the original equations get modified so that each endogenous variable gets expressed as a function of the set of exogenous variables and, possibly, the errors.

Reduced sample: A term which has been sometimes applied to the involuntary censoring of a sample when some items have been no longer observation.

Reduced variate: Refers to a standardised variate which is using location and scale statistics other than the mean and the standard deviation.

Reference period: In one sense this term has been synonymous with base period. It may also refer to the length of time, e.g., week or year for which data has been collected.

Reference set: An alternative term for fundamental probability

Refusal rate: In a sample survey we attempt to obtain information from all eligible selected sample elements. Some of these may refuse to provide the information. The proportion of those eligible elements, who refuse is called the refusal rate.

Refusals Refer to one of the categories of no response those who although successfully contracted, refuse to provide the information.

Regressand A synonym for dependent variable in a regression relation. This term is sometimes used instead of dependent variable in regression analysis.

Regression coefficient: Is the coefficient or any regressor in a regression equation. Thus, if the regression equation is simply $Y= a + bX + \varepsilon$ the regression coefficient is b.

Regression curvilinear: A relationship between two for more variables in which there is not a constant ratio of change between them.

Rejectable quantity level: Refers to the level of quality as determined by percentage defective, for example, for which a buyer would wish to have only a low probability of accepting. This probability has been the consumers risk.

Rejection: In statistics the term rejection is applied to the rejection either of a hypothesis, or in the case of quality control of a set of items of a given quality.

Rejection region in the theory of testing, hypothesis this term refers to a region of the sample space such that if a sample point falls within it the hypothesis under test get rejected.

Rejective sampling: Refers to sampling with unequal probabilities with replacement in which the whole sample has been rejected as soon as any individual has been selected a second time.

Relative cyclical residual: A measure of cyclical variation, it uses the percentage deviation from the trend for each value in the series.

Renewal Process Refers to a class of stochastic processes in which unit results in its replacement.

Repeated survey: Refers to a sample survey which is carried out more than once with essentially the same questionnaire or schedule but not necessarily with the same sample units.

Replacement cost: The current cost of materials and boughtout parts and components; the expected or forecast cost of replacing a wasting fixed asset on life expiry. Both have an important part to play in the calculation of company net profits, especially in periods of inflation and deflation.

Replacement process: Refers to a sequential control process for which, at various points, the system can be returned to some initial state.

Residual heading: A heading in a statistical classification designed to render the classification, or a section of it, exhaustive.

Residual method: A method of describing the cyclical component of a time series.

Residual treatment effect: In experiments which get continued over several consecutive periods of time on the same individual it has been important to consider 'whether the effect of the experimental treatments administered during one period has been carried over into the next and subsequent periods.

Residual variance: Refers to that part of the variance of a set or data which remains after the effect of certain systematic elements such as treatments is removed. It is able to measure the variability due to unexplained causes or experimental error.

Response metameter: Refers to the transformed measurement of the response to a given stimulus.

Response rate: In a sample survey an attempt is made to obtain information from all eligible selected sample elements.

Those eligible elements who provide the information are the respondents. The proportion of all those eligible elements in the sample who do respond is called the response rate.

Response variance Refers to a measure of the variability of those contributions to response error that tend to cancel out with sufficiently large samples.

Restricted randomisation: In complex factorial designs the randomisation of the allocation of treatments could not wholly eliminate systematic features which have been felt to be undesirable. The situation may sometime be met by imposing conditions of the randomisation which is they considered to be restricted.

Retail-price index: Formerly the costliving index; compiled monthly and published monthly through the press.

Return time: In time series, this term refers to time taken by the series to return to some assigned value, usually an extreme value, as, for instance, the return period of flooding in a river.

Return state: In the theory of Markov chains, this state refers to a state which communicates with itself. A state which communicates with no state, not even itself, is termed as a nonreturn state.

Revenue productivity index: A productivity index that rejects value added as the measure of output, and adopts instead a production index divided by the price index of the company's output weighed by reference to value in the base period.

Rho: The Greek letter (written ρ) usually used to symbolize the rank difference coefficient of correlation.

Riemann distribution: Zipf postulated the Riemann distribution

$$\phi(R) = [\xi(s)^{-1} R^{-1}$$

where $s>1$, R = 1, 2, 3

as a suitable mathematical representation of word frequency distributions.

Right and wrong cases method: Refers to a method of analysis which was proposed by Muller for quantal data arising from psychophysical experiment.

Right angular design: Refers to an association scheme for partially balanced incomplete block designs and defined four associate class designs.

Risk function: The word 'risk' finds use in actuarial statistics has found one special use in the theory of decision functions.

Robust: A test procedure is robust if it is insensitive to departure from one general assumption on which it is based.

Robustness: Several test procedure involving probability level dependent for their exactitude on assumptions concerning the generating mechanism, e.g., that the parent variation is Normal.

Root Estimator (of a mean) When we do sampling from a highly skewed population, the root estimator, (1–c) $\overline{x}$+cu^2 where $u_i = \sqrt{x_i}$, can give substantial gain in efficiency over $\overline{x}$ as an estimator of the population mean.

Root mean square-deviation: Refers to the square root of the second moment of a set of observations which are taken about in, that is to say, the square root of the some arbitrary origin, that is to say the square root of the mean square deviation or mean square error.

Root mean square error: Refers to an alternative name for root mean square deviation.

Rose diagram: Refers to a form of diagrammatic representation for grouped angular data.

Rosenbaum's test: Refers to a noparametric test for the equality of the scale parameters of two populations known to have equal medians.

Rotatable design: A *k*-actor experiment may be defined by Box & Hunter to be having a *d*th order rotatable design if the expected yield has been a polynomial of degree and in the factor levels and the variance of the estimated response function of the distance from the centre of the design.

Rotation (rotazione): There has been said to be functional rotation (rotazione funzionale) if the role of dependent and independent variables get interchanged; e.g., from $y = f(x)$ to $x = f(y)$.

Rounding: Adjusting a number, up or down to a needed level of accuracy. There have been several ways in which the level of accuracy may be presented, such as to a given number of decimal places, to a given number of significant figures, to the nearest whole number to the nearest hundred, thousand, etc.

Rounding error: Refers to the error which get resulted from the process of rounding, i.e., by omitting some end digits and replacing by zeros if required.

'Round robin' design: In the case where the same *n* objects ate to be used repeatedly throughout an experiment we are having a special form of resolvable design which may or may not be cyclic.

Route sampling: A procedure Which is similar to line sampling and used in surveys of crop acreage in districts which have been well provided with roads.

Row percentage: Refers to the percentage of any particular row in a given cell, For example, if the row total is 400 and a particular cell value is 64, its row percentage is 16%.

R-technique: A statistical technique commonly used in factor analysis in which the inter-correlations among a number of tests administrated to the same subjects are arranged in a matrix.

Run: In statistics, a series of identical symbols either numerical or nonnumerical which may or may not be preceded or followed by different symbols.

Runs test: A nonparametric statistical test that may be used with a single sample to test whether that sample is random, that is whether the case included represent a random sample of population from which they where drawn.

S

Sack's theorem: A theorem given by Sacks. This theorem states under very general conditions that when the RobbinsMunro process has been used with constants $a_n = c/n$ (c a constant), the distribution of $(x_n - \theta)$ is asymptotically Normal with zero mean.

Sample: In statistics, a Portion of the total number of cases having a given characteristic. A sample consists of a limited number of cases selected for study from a particular population.

Sample, accidental: A sample selected purely on the basis of convenience and accessibility without regard to the extent to which it is representative of the population from which it is drawn.

Sample, area: A type of sample selected on the basis of geographic areas. The total region to be surveyed is divided into geographic areas from which a designated number of areas are randomly chosen, These in turn are subdivided into smaller areas from which a designated number again are randomly selected. Within the finally selected subareas, all or a specified proportion of the households are studied.

Sample, biased: A sample selected in such a manner that certain types of cases have a greater chance of being included than other types of cases. Thus the sample is not representative of the population from which it was drawn, but differs from the population in some systematic fashion.

Sample census: A slightly unsatisfactory term. The word census has two meanings : (1) a complete enumeration of a population and (2) an enumeration in which only basic or straightforward data are collected.

Sample cluster: A sample selected by using clusters or groupings of cases rather than selecting individual cases directly from the total population.

Sample design: Sample design has two components the selection process and the estimation process. The selection process consists of the rules, procedures, and operations by which some members of the population are included in the sample. The estimation process consists of the rules and operations for computing sample estimates of the population parameters or population values.

Sample size: The number of units included a sample. It generally refers to the number of elements in the sample. In multistage sampling, for example, the sample size has been the number of final stage units in the sample. Samples are often called small if their size is under about 30.

Sample space: Refers to the set of individuals or items available to be sampled; the parent population from which a sample has been drawn.

Sample statistic: A term which is synonymous with statistic any measure calculated on the basis of the observations in the sample.

Sam stratified: A sample selected by dividing a, population into categories in on the basis of certain relavant variables and selecting a determined number or cases from each category, thus assuring that each category will be represented in the sample.

Sample survey: The term used for a survey which is curied out using a sampling method, i,e., in which a portion only, and not the whole population, is surveyed.

Sample, systematic: A sample obtained case from an arbitrary list of all the cases in the population.

SAS: A computer program for doing regression and other statistical analyses. Other commonly available packages include MINITAB and SPSS.

Saturated model: If, in a factorial experiment, q_i denotes the *i*th factor combination, binomial probability of success of success for the ith model with i parameters would be termed a saturated model.

Scalar: A number as distinct from a vector or matrix.

Scale: A standard of measurement consisting of a graduated arrangement of units.

Scale point: A particular point on a scale.

Scales of measurement: When a numerical mark or code is allocated to an individual for a particular characteristic, it is necessary to be able to define the meaning of this code in relation to the scores of other individuals in the population.

Scale parameter: Refers to a parameter of a frequency

distribution which gets functionally related to the scale of the, variable, e.g., the standard deviation in a normal distribution.

Scatter: A visual indication of the correlation on existing between two series of statistics. The extent to which the does in a scatter diagram fail to fall into alignment.

Scatter coefficient: A term given by Frisch to indicate a property of a multivariate distribution. It has been the square root or the determine ant whose elements are the intercorrelations r_{ik} between the pairs of variance, that is to say has been the square root of the correlation determinant.

Scatter diagram: A diagram showing the joint variation of two variates x and y. Each member is represented by a point whose coordinates on ordinary rectangular axes are the values of the variates which it bears. A set of *n* observations thus provides n points on the diagram and the scatter or clustering of the points exhibits the relationship between x and y.

Scatter diagram: A diagram showing a cluster of points representing observations of corresponding values of two variates, such as x and y, such that the two corresponding observed values are co-ordinates of each point on the corresponding; axes.

Scedasticity: A term, not commonly used, which denotes dispersion. It arises now only in the terms homoscedasticity and heteroscedasticity

Scenario writing: A technique that attempts to set up a logical sequence of events in order to show how, starting from the present situation, a future state may evolve step by step. The purpose, however, is not to predict the future

Schach's two-sample tests: Refers to a class of two-sample nonparametric tests for the equality of circular populations, which was proposed by Schach.

Scheffe's test: It is a conservative method, which is used for testing the significance of one or more comparisons of mean values arising in analysis of variance-where the comparisons, get selected, by inspection as being of interest.

Schuster periodogram: Refers to an alternative name for the unqualified term 'periodogram' so-called because it was introduced by Sir Arthur Schuster

Schvrarz's inequality: Refers to an inequality which in its probabilistic form states that for two arbitrary random variables A and Y

$$\{\zeta(XY)\}^2 < \zeta(X^2)\zeta(Y^2)$$

whenever the mathematical

expectations exist. A generalisation of this result for non-negative random variables has been Holder's inequality.

Score: A rating of an individual's performance in quantitative terms or an assessment of some characteristic of an individual on a quantitative scale.

Score, normative: A score expressed in terms of the population distribution.

Score, raw: A numerical score in the original form that is obtained from a test.

Scoring, method of: An iterative process which is useful for solving non-linear maximum likelihood equations.

Screening design: Refers to the statistical design for a programme of experiments which has the object of selecting a promising subset of treatments for further development.

Seasonal variation: Periodic behaviour in a time series whose period is exactly a year is referred to as seasonal variation. It is caused by climatic variation over the year or by some other phenomenon linked to the annual calendar.

Applications Making seasonal corrections to data such as those used in calculating business ratios, and in budgeting.

Second limit theorem: It is a theorem which, states that if the moments of a sequence of distribution functions Fn tend to the moments of a distribution function F, moments of all orders existing, then F_n tends to F, provided that the latter has been uniquely determined by its moments.

Secondary data: D^ta which have been removed at least one stage from source. They are structured or processed in some way.

Secondary process: In many types or stochastic process it has been possible that the events or occurrences constituting the process may also be characterised by one or more additional random variables. Each of these may be considered as a secondary process.

Secondary unit: In sampling, a synonym of second stage unit.

Second-degree equation: A mathematical form used to describe a parabolic curve that may be used in time series analysis.

Secular trend: Refers to an alternative name for trend in time series which is sometimes reserved for a trend extending over a long period of years, say, centuries, as against 'trends' extending over decades.

Segmented regression: Segmented regression has been used when a complete curve to be fitted consists of two or more submodels, and those have to be joined at points whose abscissae are to be estimated.

Selection bias: There is selection bias in a procedure for sample selection which systematically discriminates against some part of the population and gives either no chance of selection or an unknown chance of selection to that section of the population.

Selection with equal probabilities: A sample design in which all the elements in the population or stratum under consideration are allocated the same probability of being selected into the sample.

Self-avoiding random walks: Refers to a random walk in which the units are having physical size and no two units may occupy the same region of space ; or, on a lattice, a walk which at no point crosses or touches itself.

Self- correlation coefficient Refers to an alternative term for a reliability coefficient, but one which is better avoided.

Self-conjugate latin square: Refers to a Latin square which remains the same if its rows and columns have been interchanged, i.e., it is symmetrical about its main diagonal.

Self-weighting estimator: Refers-to an estimator which, for the sample design' being used, gives equal weights to the sample observations.

Semi-average, method of: It is a particular case of the method of selected points in which the data have been divided into two equal groups and a straight-line drawn through the mean of the groups or two other representative points, one in each group. This method has been used to provide a rapid estimate of a linear regression line.

Semi-averages, method of: A method of estimating trend. The data are in the form of a time series; the figures are divided chronologically into two groups, and the means of each taken, These means are then used to estimate the trend,

Semi-interquartile range: Refers to an alternative name for quartile deviation that is to say, one half of the distance between the two quartiles of a sample or a distribution.

Semi-invariant: In older usage this term related to what are now called cumulants. The words semi-invariant or "seminvariant' have been now better confined to statistics which are independent of the origin and are multiplied by a scale factor under transformations of scale.

Semi-latin square: Refers to an experiment, design for 2k treatments arranged in the form of a rectangle with *k* rows of *2k* ;

each row being an arrangement of the 2k treatments, each pair columns 1, 2, 3, 4, etc., being also an arrangement of the *2k* treatments ; no row or column therefore containing the same treatment more than once. It may also be considered as a k ´ *k* Latin square with each plot split. The design has been criticised on the grounds that it leads to biassed estimates of error.

Sequential test: Refers to a test of significance for a statistical hypothesis which is performed by using the methods of sequential analysis. An example has been the sequential probability 'ratio test.

Sequential testing: A technique of quality control used where testing is expensive, eg., where it involves destruction in testing for length of life.

Serial cluster: Refers to a type of cluster which is used in India in which the actual demarcation of a cluster of listing of units constituting a cluster gets avoided by means of a rule which makes use or the serial numbers already assigned to the units in the frame.

Serial correlation: The correlation between members. If a time series and members of a corresponding time series lagged by a fixed interval of time.

Series, queue: Refers to a queuing system wherein each arrival unit is served in turn by facilities 1, 2,.........k which are parts of one system. A somewhat different concept has been ex pressed by the term 'tandem queues'.

Set: Refers to a collection of numbers, objects, etc., that usually have at least one common property or characteristic. The members of a set are termed as elements.

Set-theoretic model: The model of the set used as the basis for the mathematical analysis of data. In essence, use of this model means that the data are divided into sets or categories and subcategories on the basis of some classification.

Shape parameter: A term used earlier to denote a parameter of a frequency distribution which is associated with skewness or kurtosis.

Sheppard's corrections: Refers to the calculation of moments from a grouped frequency distribution introduces certain errors because of assuming that frequencies have been concentrated at the central values of the class intervals.

Shock and error model: Refers to a system of equations which is having both stochastic elements associated with specific variables and elements associated with specific equations in the system, i.e. shocks.

Shock model: In econometric analysis, this term refers to a system of equations which is having random disturbances, as opposed to one in which the variables are subject to errors of observation.

'Short' distribution: Refers to the convolution of a Neyman type. A distribution with parameters θ, λ and a Poisson distribution with parameter ϕ.

Short term fluctuation: Refers to a fluctuation in a time series which is having a short duration; a continuing set of such fluctuations. 'Short' for this purpose has been' a somewhat arbitrarily defined expression.

Sigmoid curve: Refers to a curve which is lying between two horizontal asymptotes representing a function which increases monotonicity and has a point of inflexion somewhere near a point half-way between them; hence a curve somewhat resembling a letter. In statistical work this type of curve is met in connection with, inter alia, the distribution function of unimodal distributions, growth curves, such as the logistic, and a particular dose-response relationship in biological assay.

Significance: The term is used mainly in the context of hypothesis testing. It is said to exist when a value lies outside an acceptable interval and when, consequently, there are good reasons for rejecting a null hypothesis

Significant figures: The figures of a number that express its maginitude to a specified degree of accuracy.

Significance level: In hypothesis testing it is usual to obtain from a given set of sample data a test statistic calculated for the purpose of the test. This test statistic can only be used if its distribution under the null hypothesis is known.

Significance, statistical, The probability that an observed sample characteristic or relationship, 'or difference between samples, represents a true population characteristic, relationship, or difference and is not merely due to chance sampling variation.

Sign test: A nonparametric statistical test that may be used when data are in the form of paired observations divided into a dichotomized ranking, that is, divided into plus and minus categories indicating more and less or higher and lower.

Similar action: The name assigned to the action of mixtures of stimuli, e.g., the toxic effect of a mixture of poisons, when the stimuli have been statistically independent and additive. The effect of a mixture is then predictable from the relative proportions of the constituents and the known response of each.

Similarity, coefficient of: Any measure used to describe the closeness or similarity of elements, groups or variables. The term is used particularly in cluster analysis. The most commonly used to coefficient of similarity is the productmoment correlation coefficient.

Similar regions: In the theory of testing hypotheses a region in the sample space has been said to be similar to another if a correspondence can be set Lip between them such that the Probability of a sample point falling in a part of one is Proportional to the probability that a sample point falls into the corresponding part of the other.

Similarity index: If two individuals each bear the value of p(0, 1) variables, and in m cases they both exhibit, or do not exhibit the same variate value, the ratio m/p is a similarity index. The complementary quantity 1-(m/p) has been a dissimilarity index, and has been used particularly in Cluster Analysis as a distance function.

Similarity matrix: Refers to a matrix which presents the coefficients of similarity for each pair of elements in the data set.

Simple sample: A random sample is said to be simple when the probabilities of selection of members have been all equal and have been constant throughout the drawing.

Simple table: A table which is showing only classifications according to one variate or at the most two variates.

Simplex designs: A simplex has been the n-dimensional analogue of the triangle in, two, or the tetrahedron in three, dimensions.

The superposition of two or more simplexes to give certain kinds of rotatable designs is known as a simplex sum design.

Simplex method: Refers to a method of solving a mathematical program by the process of iterative application of rules associated with matrix inversion until an optimal solution gets reached.

Simplex technique: The general linear programming technique

Simulation: The representation of one system by means of another. In particular the representation of physical phenomena by computers, other equipment or models, to facilitate the study of such systems or phenomena, to train operators, etc.

Simulation model: Certain types of dynamic model problems cannot be solved by ordinary analytical methods because conditions change rapidly or because of the large number 'of variables. In such circumstances, it is possible that the problem can be approached by assuming the likely effects of the conditions and producing a computer model which will show, the effects of changing one or more of the variables. For example, computer models of business firms can be constructed to simulate the likely effects of trade recessions, seasonal fluctuations, theft, weather conditions, alteration of interest rates, etc.

Simulator: Refers to a physical systems which is analogous to a model under study as, for example, an electric network in which the elements have been in correspondence with those of an economic model. The variables of interest in the model appear as physical variables such as voltages and currents and could be studied by an examination of the physical variables in the simulator.

Single linkage: Refers to a form of cluster analysis. This principle on which it operates is as follows. First, the two elements with the highest similarity in the set have been linked together. The next largest value in the matrix identifies the next pair to be linked. The principle used in the process has been that an element is linked to a cluster if it is sufficiently similar to any of the elements in the cluster. This is why the method is sometimes referred to as nearest neighbour analysis-an element is included in the cluster to which its nearest neighbour belongs. The process continues until the required number of' clusters is formed.

Single sampling: In evaluating the quality of a product or material by inspecting a part rather the whole, single sampling refers to the method in which the decision whether to accept or reject the batch is taken after the inspection of a single sample. This may be contrasted with sequential sampling, where successive samples 'elements or groups of elements are drawn until sufficient evidence is available to make a decision.

Sirgle sampling scheme: In quality control, a sampling scheme where only one sample per batch gets inspected. If the number of defectives found is not more than the acceptance number, the batch is accepted. Otherwise it is rejected, or subjected to 100% inspection.

Single sampling plan: In quality control, this term refers to a procedure which provides for the drawing of one sample, on the basis of which a lot is accepted or rejected, as distinct from a double sampling plan, which may allow for indecision on the first sample to get resolved by a further sample.

Single tail test: An alternative term for a one sided test.

Size of a region: In the theory of testing statistical hypothesis, the size of a critical region has been a measure of probability and has been the same as probability of an a-arror or error of the first kind. For composite hypotheses it is sometimes used to denote the limits of the a-error where no similar regions exist.

Size (of a test): The size of the test of a statistical hypothesis (H_0) refers the probability (a)that the hypothesis gets rejected when it is true.

Skew regression: An obsolete term for curvilinear regression.

Skewness: A form of asymmetry in a unimodal frequency d distribution. The distribution is positively skewed when more than 50% of elements lie below the mean and negatively skewed when more than 50% of elements lie above the mean

Slippage test: Refers to a significance test of k samples in which the hypothesis has been one of homogeneity in the means, a against the alternative that one member or set of members ha slipped' away from the others.

Slope: A constant for any given straight line, whose value represents how much each unit change of the independent variable changes the dependent variable.

Slope ratio assay: Refers to a general class of biological assay where the dose response lines for the standard test stimuli have been not in the form of two parallel regression lines but of two lines with different slopes intersecting the ordirate corresponding to zero dose of the stimuli.

Slutzky process: A synonym for the moving average process.

Slutzky-yule effect: Refers to an effect in the averaging of random series which was studied independently by Slutzky and Yule.

Small numbers, law of: A term introduced by von Bortkiewicz to describe the behaviour of rare events obeying a poisson distribution.

Small sample: A sample from which, because of its small size, conclusions have to be treated with particular caution. Hypothesis testing may involve the use of special tests.

Smirnov-birnbaurn-tingey distribution: Refers to the probability distribution Pn(ε) that if an empirical distribution fu Ction $F_n(x)$ has been constructed from a random sample of size *n* fiom a continuous probability density having distribution function F(x) then F(x) will be nowhere greater than Fn(x)+ε.

Smooth test: Refers to a test of goodness of fit between data hypothesis in which the alternate hypotheses have been considered as moving away from the null hypothesis smoothly in the parameters, i,e., with high continuity and differentiability in them.

Smoothing: Refers to the process of removing fluctuations in an ordered series so that the result shall be 'smooth-' in the sense that the first differences have been regular and higher order differences small

Snedecor's check: An approximate method of checking Standard deviation calculations in normal cases. Briefly, it involves dividing the range by 2 if n is less than 7; by 3 if it lies between 7 and 20; by 4 if lies between 21 and 40; by 4.5 if it lies between 41 and 80 by 5 if n approximates 100; and so on.

Snedecor's F-distribution Refers to a popular term for the variance-ratio distribution, so-called by Snedecor after the initial of R.A. Fisher's surname.

Snowball sample: A sample in which additional elements are generated by the units first obtained in the sample.

Snowball Sampling: Refers to a form of sampling which was proposed by Goodman whereby a random sample of *n* individuals is drawn from a finite population. Each individual has been asked to name *k* further individuals : and so on for 8-stages. The initial sample solution and its size govern the kind of analysis and inferences that could be drawn from the data.

Sojourn-time: The length of uninterrupted time that astochastic system spends in a given state.

Spatial point process: Refers to a stochastic point process in two dimensions usually represented on a plane average or smoothed series.

Spatial systematic sample Refers to a systematic sample taken in two dimensions; also termed 'plane sampling' by Quenouille.

Spearman estimator: Refers to the estimator, non-parametric, of the mean effective dose which is relevant to the Spearman, Karbar Method of bio-assay or other stimulus experiment.

Spearman's rho (r) A rank correlation coefficient defined as, the product moment correlation coefficient between two sets of rankings of a collection of objects or individuals,

Spectral average: Refers to the average over a spectral density function of the form

$$J(A) = \int A(w) f(w) dw$$

where A(w) has been a suitably chosen function. It is found, particularly in the smoothing of the spectral density function and A(w) is sometimes termed as the spectral window Special forms of such a window are known by the names of Daniel, Bartlett, Blackman-Tukey and Parzen.

Spectral density: Refers to the derivate of the spectral function.

Spectral function: A necessary and sufficient condition for $\rho(t)$, τ =0,1, 2......... to be an autocorrelation function of a discrete stationary stochastic process has been that it has been expressible in the form

$$\rho(\tau) = 1/\pi \int_0^\pi \cos\tau w dF\ (w)$$

where $F(\omega)$ denotes a non-decreasing function called the spectral distribution function with $F(0) = 0$, $F(\pi) = \pi$

Spectral weight function: A weigh function which is used in the estimation of the spectral density : individual specifications are proposed, for example, Daniel, Bartlett, Blackman & Tukey and Parzen.

Spencer's 15-point moving average: Refers to an example of moving average having unequal weights. It has been used to smooth or extract the trend from a time series. The smoothed series at time t is a weighted average of past and future observations with weights that are symmetric. Algebraically a symmetric moving average of length 15 can be written

$$S_t = \sum_{j=-7}^{+7} aj\ x_{t-1}$$

Spurious correlation: The introduction into a correlation or regression analysis of a theoretically irrelevant factor, or the introduction of two or more factors as independent variables which, in effect, provide measures of the same force.

Square root transformation: Refers to a variate transformation which has Peen used to 'stabilise the variance' of sample data drawn fro. Poisson populations ; that is to say, to give variates whose variance would be nearly independent of their means.

Squariance: A term which was proposed by Pitman in place of the phrase 'sum of squares about the mean' for the sake of brevity.

SRS: A commonly used abbreviation for simple random sampling.

Stability test: Refers to a rough but convenient test for binomial variation in data which accrue over a period. The cumulative proportion of successes has been plotted against the number of observations. If the variation has been binomial the graph of successive points 'settles down' to a straight line, or nearly so, with diminishing fluctuation about such a line.

Stabilisation of variance Refers to

the process of transforming a variate whose distribution has been dependent on a parameter, so as to make the variance of the transformed variate as insensitive as possible to the values of the parameter.

Stable process: Refers to an alternative term for stationary process. A distribution law has been said to be stable if the convolution of two variates which follow it has been also distributed in the same form : e.g., the sum of two independent Normal variates has been Normal and hence the Normal law has been stable in the sense.

Stable state: Refers to a state of a Markov chain in continuous time for which the expected sojourn time has been greater- than zero.

Staircase distribution: Refers to a picturesque but unnecessary name for a discontinuous distribution which is based upon the fact that its distribution function is a step function.

Staircase method: Refers to an alternative name for the 'up and down' method especially when used in connection with fatigue tests.

Standard score: Refers to an alternative name for z-score.

Standardised deviate: Refers to the value of a deviate which is reduced to standardised form by subtracting the parent mean and then dividing by the parent standard deviation.

Standardised mortality ratio Refers to an index number in the Paasche form used in the analysis of vital statistics. The ratios of the age specific death rates for the given year to similar rates for the base or standard year have been weighted by the expected' deals in the given year.

Standardised variate: Refers to a variate in standard measure.

State of nature: Future event not under the control of the decision maker.

Stationary distribution: A phrase which has been sometimes used to denote a distribution, e.g of a human population which remains constant in time. There has been some danger of confusion with a stationary stochastic process and perhaps the expression has been better avoided.

Stationary process A stochastic process is said to be stationary if its statistical properties are not changing over time.

Statistic: A mathematical value that summarizes a characteristic of a sample, A statistic may be a measure of central tendency, of dispersion of correlation, of a difference between two samples, or any other summarizing measure based upon, sample data. It is used to estimate a parameter of the population from which the sample was drawn, and is sometimes referred to as a parameter estimate. Since most social research deals with samples rather than entire populations, most research yields a statistic rather than a parameter.

Statistical dependence: The condition when the probability of some events is dependent upon, or affected by, the occurrence e other event.

Statistical independence: The condition when the occurrence of one event has no effect upon the probability of occurance of any other event.

Statistical inference: The process of inferring generalizations about a population on the basis of sample data. Statistical inference includes the estimation of population parameter using the testing of statistical hypotheses about populations using data obtained from samples. Statistical interface is based upon probability theory.

Statistical model: A set of assumptions underlying a statistical test. Every statistical test assumes that certain conditions are met. Usually these assumptions refer to the form of the data, the nature of the sample, the nature of the population from which the sample was drawn, and the chapter of the variables being studied. The test is valid only if these conditions are met, and thus it is important to know the statistical model on which a lest is based before that test, is, used.

Statistical progression: A time series with a seasonal pattern or a secular trend or both. A progressive average is the AM based on a cumulative total. It is not a moving average.

Statistical schedule: A device which sets out the relevant figures of a dependent variable for ranges of the independent variables often based on a regression equation.

Statistical table: A systematic arrangement of numerical data into a series of rows and columns in order to facilitate the comprehension and comparison of the data. The presentation of data in a statistical table involves the classification of the data, into the categorites or class intervals specified in the column and row headings.

Statistical test: A procedure used with quantitative sample data to estimate the probable truth of a hypothesis about the population from which the sample was drawn, The formulas used are based on certain principles, assumptions, and rules of mathematical logic. A wide range of data.

Statistical tolerance interval Refers to the interval between an upper and a lower statistical tolerance limit.

Step down procedure: Refers to a procedure whereby differences have been examined in one observed variable first and then in another observed variable after eliminating by regression the effect of the first, and so on. The variables have been in some sense, arranged in a descending order of importance.

Stephan's iterative process Refers to an iterative process which was proposed by Stephan, for adjusting sample frequency tables when, the expected marginal totals are known.

Stepwise multiple regression: A method of eliminating or including variables in regression analysis by means of a stepwise procedure.

Stepwise regression: Refers to a method of selecting the 'best set of regressor variables for a regression equation. It is carried out or by introducing the variables one at a time or by beginning with the whole set and rejecting them one at a time.

Stochastic convergence: Refers to one of several concepts in probability. A sequance $\{x_n\}$ of variate is regarded to converge stochastically to a variate x if

$$\lim_{n \to \infty} \text{pr}\{|xn - x| > \varepsilon\} = 0$$

for every $\varepsilon > 0$. This is also termed as convergence in probability and convergence in, measure.

If x_n and y_n have been variates and, as n tends to infinity

$$\Pr\{|x_n - y_n| > \varepsilon\}$$

tends to zero x_n is considered to converge stochastically to y_n but by convention the case where x_n. and y_n do not converge has been usually excluded.

Stochastic dependence: Refers to the reoationship between variates which are not independent as contrasted with mathematical dependence, which has been a relationship between variables. In statistical usage either form has been often referred to as 'dependence', the meaning generally being clear from the context.

Survey: A study or investigation of a population, usually human beings or economic, social or political institutions.

Survey method: All methods, procedures and techniques which are used in carrying out social surveys are subsumed in the term survey methods, Thus the term is covering questionaire design, sample design, methods of data collection and methods of data analysis.

Survey population: The population actually covered by a sample survey. Idealy the survey population' should be identical to the target population for which results are desired, but, mainly due to non-coverage and non-response, there may be differences between the two. Other modifications may be—necessary simply to make the survey practicable.

Survey sampling: It deals with methods of selecting a sample from a population so as to use the sample observations, together with information on the method of sampling, to make references about the whole population.

System: The term system is generally used for something which has in it an element of design, which can be represented by a model. Thus a model is a representation of reality, whereas a system comprises that set of states or processes in reality which the model is designed to represent.

Systematic sample: A sample obtained by the method of systematic sampling. The term is usually confined to samples drawn by selecting units at equally spaced intervals-the sampling interval. The first sample unit is selected from the first sampling interval by some random method.

Systematic sampling: Refers to a sampling in which each member of the sample is chosen by ordering the frame in some way, and then selecting individuals at regular intervals. Thus a 1% sample of telephone subscribers could be taken by selecting the 11st, 101st, 201st of the names listed in the telephone directory.

T

Table (Tabulation): It is a systematic summary presentation of data. Tables maybe:

1. Single-column univariate frequency tables, and single column category tables showing the frequencies or relative frequencies of a number of dissimilar categories.
2. Bivariate tables.
3. Statistical tables giving critical values of a statistic, such as t, F, χ, ρ, etc. This usage has been very common.

Table twisting: A tabulating process designed to present figures in the most effective way, either as information, or in furtherance of an argument. The process consists of twisting the table so that the stub becomes the column headings, and vice versa.

Tabular presentation: The systematic presentation of statistics.

Tail area: Refers to the portion of the area under a frequency curve which is lying between the start of the distribution and some ordinate lying between the start and the mode; or symmetrically, between some ordinate lying between the mode and the end of the variate range and the end of the distribution. The term is applicable only to distributions which 'tall off' at their extremes, i.e., have frequency functions tending to zero.

Target population: Refers to the population about which we wish to get information. Every population must be defined in terms of (1) content, (2) units, (3) extent and (4) time.

Target sample: This term is used to denote the sample originally selected or designated. Non-response may make the achieved sample to fall short of the target sample.

Telescoping: Refers to a type of memory or recall error. When respondents have been asked to provide information about the past, the question usually involves a reference period, for example, 'How much money did you spend on household repairs in the last three months? Frequently the respondent may include in his answer, by mistake, items of expenditure which occurred more than three months previously, particularly if the expenditure had been large, This process is termed as telescoping.

T-distribution: Refers to the distribution of the means of samples of given size, standardized using estimated values of the parent standard deviation. It has been used in the t-test.

Temporally continuous process: Refers to an expression, which is sometimes used to denote a stochastic process which depends upon a continuous time parameter. The terminology is not recommended.

Temporary homogeneous process: Refers to a stochastic process for which the transition probabilities have been same for any time interval of length *t*.

Terminal decision: In sampling schemes of a sequential by this terms refers to a decisions the sampling process.

Test coefficient In factor analysis, refers to a synonyms for fact loading.

Test of independence: A statistical test of proportions or frequencies in determine if membership in categories of one variable different as a function of membership in the categories of second variable.

Test of normality: Refers to a test of a set of observations see whether they could have arisen by random sampling from Normal population. Such tests may be done by a comparison of the sample distribution function with a Normal distribution function.

Test statistic: 1. A statistic which has been calculated from a sample in order to test a hypothesis about the parent population from which the sample is drawn; examples are X^2, and t. 2. A statistic calculated from more than one sample (like F, U etc.) to compare the parent populations from which they are drawn.

Theoretical frequencies: Refers to the frequencies which would fall into assigned ranges of the variate if some theoretical distribution law had been exactly followed, as distinct from the actual frequencies which may be observed in a sample.

Theoretical variable: This variable is used to denote a variable or variate introduced into structural relations but not directly observable; as distinct from a 'true' variable or variate, which has been observable except for errors of observation and an observable' variable or variate which could be directly observed.

Theory of runs: A theory developed to allow us to test samples for the randomness of their order.

The expected value of perfect information: The difference between expected profit (under conditions of risk) and expected profit with perfect information.

Three dimensional lattice: Refers to a general class of lattice design of which the cubic lattice has been a particular case. For example, 120 treatments could be laid out as a 4 × 5 × 6 three- dimensional lattice. This would be considered as equivalent to a factorial design having three factors at 4 5 and 6 levels.

Three point assay: Refers to the simplest case of the general class of $(2k + 1)$ point slope ratio assays. This design has been useful only where the validity of the assay has been known a *priori*, since in the analysis of the results no degree of freedom have been available for tests of validity.

Three-quarters High Rule Refers to an empirical rule for ensuring that a bar chart provides the right impression. It states that the maximum height should be three-quarters of the horizontal displacement.

Three-stage least squares: Refers to a method for estimating the coefficients in a simultaneous equations model as a whole than as a set of single equations in isolation.

Three-stage sample: Refers to a special case of multistage sampling. The population is assumed to be classified into first-stage units (primary sampling units), Thus the full procedure would, be (1) to select a sample of constituencies the first stage. (2) to select a sample of polling districts within each selected constituency of the second stage; (3) to select a sample of individuals within each selected polling district-the third stage. The method can be applied to any population in which suitable units are available. In a sample of industrial workers, for first stage units might be firms, the second stage units departments within firms; and the third stage units the individual workers.

Ticket sampling: Refers to a method of selecting a random sample in which the characteristics of each member of the population have been noted on a separate card and the required number of cards are drawn from the resulting pack.

Tied ranks: In ranking procedures, two or more objects or persons, may achieve levels of success (or failure) which are indistinguishable from each other. In this case, it is usual, when calculating rank correlation coefficients, to allot ranks to each equally successful constituent of the group equivalent to the mean of the ranks which they would have received if they were not equally successful.

Time antithesis: Refers to an index number formula which was derived from another formula by interchanging the subscripts denoting the base period and the given period and then taking the reciprocal.

Time sampling: Refers to the sampling of observations of a variate over a period of time, as distinct from the extraction of a sample from a population at a given point of time. Contrast, for example, the sampling of temperature measurements at a particular place on each day for a number of years (time sampling) with the sampling of weights of children at a school on a given day.

Time sequence: Refers to a set or array of variate values which have been ordered chronologically (i.e., in time order) as distinct from a time series, where a variate has been observed consistently at fixed intervals of time.

Tolerance distribution: Refers to the distribution among a number of individuals of the critical level of intensity at which a stimulus will just produce a reaction in each individual. Although the distribution of these tolerances can be skew, it has been often possible to make it approximately Normal by a simple transformation like the logarithmic.

Tolerance factor: In quality control, this term refers io the difference between the upper and lower tolerance limits divided by some measure of the variability of the product, usually the standard deviation. Sometimes one half of this quantity may be considered as the tolerance factor, especially where the distribution of the variate under measurement has been symmetrical.

Tolerance limits: In quality control, the limiting values between which measurements must lie if an article has to be acceptable, as distinct from confidence limits.

Tolerance number of defects Refers to an expression which in better English would be 'tolerable number of defects'. It would be obtained by multiplying the lot tolerance percent or fraction defective by the size of the lot or batch submitted for inspection.

Total correlation: The zero order correlation between two variates, i.e., the correlation between the original data rather than between residuals after some common variation has been abstracted.

Total determination coefficient of: In regression analysis, it refers to the square of the coefficients of multiple correlation: R^2 It represents the proportion of the total variance of the dependent variate which has been accounted for by the variation of the independent variables in the multiple correlation.

Total effect: The total affect of a variable on another in a path model has been the sum of the direct effect and indirect effect. Thus it represents all the influence one variable has on another whether it operates directly or through its impact on some intermediate variable in the model.

Total regression: Refers to a regression coefficient of zero order, i.e., a coefficient which is involving only one dependent and one independent variate. The expression has been probably better avoided.

Tranche: A share, literally a slice.

Transfer function: Alternatively termed as the frequency response function.

Transformations: In mathematical manipulations for converting one variable into a different form, so we can fit curves as well as lines by regression.

Transformation set of Latin squares: If the rows, columns and letters of a Latin Square have been permuted, the resulting set of Latin squares has been termed as the transformation set. In the case of squares of certain sizes, e.g., 6 × 6 squares, not all the squares of a transformation set would be different.

Transient state: Alternative name for nonrecurrent state.

Transition matrix or stochastic matrix: Refers to a matrix relating the probabilities of a stochastic process being in a particular state given its previous state. Symbol: T. It has been used to model a stochastic process. A transition matrix is also associated with every Markov chain.

Transition probability: In the theory of stochastic processes, the transition probability may be defined as the conditional probability, that a system in state E_i would be in state E_k at some designated later time.

Translation parameter: Refers to a parameter of location.

T-Ratio distribution: Refers to the distribution of the ratio of two random variables which follow the bivariate. It was developed by Press The distribution arises in Bayesian approaches to inference in econometrics. It can be shown that the distribution is having no moments.

Treatment: In experimentation, this refers to a stimulus which is applied so as to observe the effect on the experimental situation, or to compare its effect with those of other treatments. In practice, 'treatment' may refer to a physical substance, a procedure or anything which has been capable of controlled application according to the requirements of the experiment.

Treatment mean square: Refers to a mean square in a variance analysis which is assignable to differences among the effects of one of the experimental treatments.

Tree diagram: Refers to a diagram in which events or relationships have been represented by points joined by lines. It is laid out so that the rules, multiply along the branches' and 'add vertically' apply.

Trend Refers to the underlying direction and rate of change in a time series, when allowance is made for the effects of seasonal components, cycle components and irregular variation. The equation of a trend may be termed as the trend line: it need not be straight although it often is.

Trend empirical: A trend followed in a set of data for which the investigator has no logical or theoretical explanation. Such a trend is usually identified simply to eliminate its influence from consideration in the analysis of cyclical movement.

Trend fitting: Refers to tile general trend component of a time process of representing the series. A trend could be represented by a particular curve form e.g., the logistic, or by a particular form of the general class of Polynomial in time, or by a moving average.

Trend free series: This is a series in which no trend is present either because the variable concerned historically has fluctuated about a constant level, or because whatever trend was originally in the data has been removed by statistical adjustment.

Trend linear: A trend in which the rate of change remain constant. If it were plotted on a graph, it would from a straight line.

Trial: In probability theory in 'trial' has been a deliberate attempt to generate an event which is supposed to be happening under a probabilitic scheme; e.g., the tossing of a coin is a 'trial', the outcome being one of two possible events, a head or a tail.

Trilling: A technique which is used in confluence or bunch map analysis for setting out systematically all the elementary regressions which are found in all possible subsets of variables in a regression equation.

Triple comparisons: The statistical model for paired comparisons could be extended to the (t_2) triple comparisons that may be formed from t treatments. The asymptotic relative efficiency of triple comparisons, compared with paired comparisons, has been 15 and $9/(4\pi)$ when compared with ordinary analysis of variance.

Trough: Refers to an observation in a discontinuous time series which is lower than each of the two neighbouring observations; or, in the continuous case, a point where the series is having a minimum.

True mean: An alternative, although little used term for the mean of a population, as distinct from the mean of a sample.

True regression: Refers to an expression which has been sometimes used in reference to a sample to denote the regression of one variate upon another which would have been obtained if there had been no errors of observation in the independent variable, The usage has been open to misunderstanding owing to possible confusion with the 'parent' regression, as distinct from sample regression and is probably better avoided.

T score: A Z score (standard score that has been modified by multiplying it by ten and adding fifty in order to eliminate negative, zero, and decimal scores.

T test: A statistical test used to determine the probability that a statistic (summarizing value for a sample) obtained from sample data is merely a reflection of a chance variation in the sample(s) rather than a measure of a true population parameter. The *t* test is used instead of the critical ratio when the size of sample of certain size. The *t* test is used instead of the critical ratio when the size of the sample is small.

Turning point: In an ordered series, this term refers to an observation which has been a peak or a trough. When several contiguous values have been equal, and greater than or less than the neighbouring values, a convention has been required to determine which has been regarded as the turning point, e.g., the middle one may be chosen.

Two-by-two frequency table: A term which is used for the presentation in tabular form of data subject to double dichotomy.

Two-phase sampling: A sampling procedure in which the final sample is not selected directly from the population but from the first phase sample, through which information is obtained into improve either the selection or the estimation or both.

Two-sample test: Refers to a test on two different, nor-matched samples. This test is often, used to mean a test of the null hypothesis that two samples, have been drawn from Parent populations with the same mean, or the same parent population. It is this test which is the subject of the rest of this entry.

Two sided test: An alternative name for double tailed test.

Two-stage least squares: Refers to a limited-information method which can lie used to estimate the coefficients of any identifiable equation in isolation from the other equations in simultaneous equations model.

Two stage sample: Refers to a simple case of a multistage. In this case the population to be sampled has been first classified into primary units, each of which consists of a collection of the basic sampling unit, the secondary unit.

Two-stage sampling: A simple case of multistage sampling. The population is classified into groups of elements—the primary sampling units. A sample of these primary sampling units is selected—the first stage; the elected primary units are then classified into the basic sampling units-the secondary sampling units; and a sub-sample of secondary units is then selected from each—this constitutes the second stage.

Two-tailed test: A statistical test used to test a hypothesis when the direction of the difference between samples or relationship between variable is not predicted in the hypothesis.

Two way classification: Refers to the classification of a set of observations according to two criteria of classification as, for example, in double dichotomy or a correlation table.

U

U curve: In statistics, a frequency distribution curve in which the highest frequencies are at the two extreme ends of the curve and the lowest frequencies are in the middle.

Ultimate cluster: Refers to the set of elements from a particular primary sampling unit which are included in the sample.

Unadjusted moment: Refers to a moment of a frequency distribution before any adjustment has been made for the effect of grouping the observations, e.g., before the application of Sheppard's corrections.

Unbiased: A sample, estimation procedure or estimator which is -free from bias has been said to be unbiased. Thus no systematic distortion or error is present in the procedure. Such a systematic error should get distinguished from a random error, which may give rise to a distortion on any particular occasion but balances out (cancels out) on the average.

Unbiased confidence intervals: A system of confidence interval which has been selective in the sense of covering the true value of the parameter with the assigned probability l-a and has been covering other values as little as possible has been said to be unbiased if $Pr\{\delta \varepsilon \theta \mid \theta\} = 1 - a \geq Fr\{\delta \varepsilon \theta \mid \theta'\}$ where δ denotes a typical interval: θ denotes the true value and θ' an alternative, This usage of 'unbiased' has been analogous to that for tests of significance.

Unbiased error: Refers to an error which may be considered as a member drawn at random from an error population having zero mean. Thus in the long run positive and negative errors tend to, cancel out in the sense of having a mean which tends to zero.

Unbiased estimating equation Refers to an equation for the estimation of a parameter in which the terms get unbiased estimators of the corresponding parent values. It does not follow that the estimator of the parameter has been then unbiased itself. For example, if the estimator t of a parameter θ has been given by $a\text{-}bt=0$, where a and b have been variates, a and b may be unbiased and hence the equation has been unbiased, but the ratio a/b may still give a biased estimator of θ.

Unbiased estimator If an estimator is having an expected value, for all sample sizes, equal to the parameter it is designed to estimate, it is an unbiased estimator. Thus in repeated sampling, the estimator takes on average the true value of the parameter. Therefore although for any particular sample the estimate obtained may not be correct (i.e., may not be equal to the parameter being estimated), the average taken over all possible samples will be equal to the parameter. The estimator has been thus free from systematic error or distortion.

Unbiased test: A test of significance may get unbiassd in a manner different from the use of that term in estimation.

Unidimensionality: Refers to the capacity to be measured in only one dimension. A univariate frequency array, for example, has been a one- dimension array. Although two axes have been used for the purpose of showing frequency values, in practice a single variate could be depicted along a single axis or by using a single vector of frequencies.

Uniform distribution. Refers to an alternative term for the rectangular distribution.

Uniformly better estimator. In comparing two estimators T_1 and T_2 of a population parameter, if, using a particular criterion, T_1 has been at least as good as T_2 at all possible parameter values, and better at some parameter values, we say that T_1 has been uniformly better than T_2.

Uniform sampling fraction: If a sample has been selected from a Population which has been grouped into strata, in such a way that the number of units selected from each stratum has been proportional to the total number of units in that stratum, the sample has been said to have been selected with a uniform sampling fraction.

Uniform scores test: Refers to a nonparametric test of the equality of two circular populations which was proposed by Wheeler and Watson.

Uniformity trial: Refers to an experiment, or set of trials, in which each experimental unit is receiving exactly the same treatment. The object, inter alia, may be to estimate some standard characteristic of that treatment or to investigate some aspects of the experimental technique, e.g., plot size or layout in agricultural trials.

Unimodel: Refers to a frequency distribution which is having only one mode. For example, the normal distribution is having only one mode, which coincides with its mean and median, and asymmetric or skewed distributions may also have single modes which do not coincide with their means and medians.

Unit: Refers to a means of measuring or counting data for a particular purpose, or a homogeneous item in which data are either counted or measured. It may be a unit of measurement, like a kilometre, pound, dollar or hour, or a unit counted for measuring *frequency, such as a* person, a brick, a share or even an examination mark.

Unit normal variate: Refers to a variate which has been Normally distributed with zero mean and unit standard deviation, It has been often written N(x ; 0, 1) or N(0, 1).

Unitary sampling: Refers to sampling in which the ultimate units have been directly chosen, as contrasted with a multistage sampling where primary groups have been first chosen.

Univariate (univariate distribution): The term univariate simply means having only one variate (or variable). A univariate distribution has been a single distribution having only one variate, which may be either continuous or discontinuous.

Unlikelihood ratio: Refers to a type of nonsequential multiple decision rule which was proposed by Hall where the sample size has been a minimum subject to upper bounds on the probabilities of making incorrect decisions.

Unreduced designs: Refers to a particular kind of balanced incomplete block design which is derived by taking all possible sets of k treatments from the full set t, each set of k forming a block. The design gets completed by permuting the blocks and arranging the treatments randomly.

Unweighted mean: Refers to a mean of a set of observations in which no weights have been attached to them, except in the trivial sense that each has weight unity.

Unweighted means method: In variance analysis, this term refers to a simple method for the analysis of a set of results for which the subgroup frequencies have been unequal. It involves taking the mean values for each subclass and carrying out an ordinary variance analysis on those means.

Upper-tailed test: A one-tailed hypothesis test in which a sample value significantly above the hypothesized population value will lead us to reject the null hypothesis.

U-shaped distribution: A frequency distribution shaped more or loss like a letter U, though not necessarily symmetrical i.e. with the maximum frequencies at the two extremes of the range of the variate.

V

Validity: 1. In statistics, correspondence between what a measure device is supposed to measure and what is really measures.

Validation: A procedure which uses external sources independent of the investigation in order to Ascertain whether the measurement process is free from bias. Checking the responses obtained in an interview against official records is an example of a validation exercise. The term is also used to denote a check carried out to find out whether a sample is representative of the population from which it was drawn.

Value index: Refers to an index number which is formed from, the ratio of aggregate values in the given period to the aggregate values in the base 1) period. Strictly speaking it is not an number as ordinary understood but a value relative.

Value ultimate: A value or ideal that a society or subgroup in a society regards as fundamental, unchallengeable, and unchanging.

V'anity effect: Refers to a form of bias encountered in surveys of human population which can be introduced into the results through distorted responses by the individuals being questioned, The person questioned fails to provide an accurate reply but makes instead an assertion more pleasing to his personal vanity.

Variable discrete: A variable that is composed of a series of distinct and separate was or categories. All qualitative variables are discrete variables, and those quantitative variable, consisting of values that are indivisible units (such is individuals) are also discrete variables. There are no values possible in between the units of a discrete variable. The magnitude of a discrete quantitative variable increases by definite steps.

Variable error: Refers to an error

which is occurring in certain types of experiments on living subjects due to factors which vary during the course of an experiment, such as tiredness, or experience.

Variance: A measure of dispersion of statistical values that is the arithmetic mean of the squared deviations from the mean.

Variance, analysis of: A method of dividing the total variation of observations into components which can be attributed to or associated with particular sources of variation, e.g., the difference between groups or classes use in classifying the observations.

Variance component: One of the objects of variance analysis have been to split up the sum of squares of observations about their mean, into portions which could be assigned to variation about their mean into portions which could be assigned to variation between the classes or sub-classes according to which the data have been classified.

Variate difference method: Refers to a method of analysis time series which is having a systematic and a random component.

Variate transformation, Refers to the transformation of one variate into another, generally by a mathematical equation connecting them.

Variation: A measure of dispersion of statistical values that is the sum of the squared deviations from the mean. To obtain the variation, each value in a frequency distribution is subtracted from the arithmetic mean.

Vector correlation coefficient: Refers to a generalisation of the product moment correlation between two variates for the purpose of measuring the relation between a p-way vector variate and a p-way vector variate.

Vicious circle: Intercorrelation between two factors that tends to cause an undesirable growth in the strength of the factors.

W

WAGR test: It is a sequential test procedure for $H_0 : p_0 + p_1$ where the proportion has been derived from observations of a Normally, distributed *variate* greater than a given value. The designation has been derived from the names Wald, Arnold, Goldberg and Rushton.

Waiting line: It is the American equivalent of the English meaning of the 'English noun 'queue'.

Waiting time: In the theory of queues this term has been self explanatory, but, in renewal theory residual waiting time has been the period from a given point in time (t) to the next renewal point, Spent waiting time has been the time elapsed since the last renewal The sum of the two waiting times has been the length of recurrence interval. In the theory of random walks residual waiting time has been termed 'point of first entry' or 'hitting point' for the interval t, ∞.

Wald-Wolfowitz run test: It is a nonparametric test of the identity of the distribution functions of two continuous populations against general alternative hypotheses, which was proposed by Wald & Wolfowitz. A sample of n independent values from the first population gets combined with a sample of n independent values from the second population, the combined sample gets ranked, and the test statistic is computed as the number of runs, where a run may be defined to be a succession of adjacently- Tanked values from the same population which are followed and preceded in rank by values from the other population or no values at all.

Walker probability function: A function which was derived by Sir Gilbert Walker in connection with tests of signi*ficance for* the ordinates of a periodogram. Developing a result due to Schuster Walker stated that the probability that one value of the intensity (S^2) of the m = ½n

independent values in a Fourier sequance of intensities, shall not get exeeded $4\sigma^2k/n$ is $1-(1 - e^{-k})^m$. The value of this function for arguments k and m has been known as the Walker function.

Weight: Refers to a number by which a value is multiplied so as to give it a required measure of importance in relationship to other values. The term is used in simple statistics in the calculation of index numbers and averages and in sampling.

Weighted average: An average where each item is multiplied by a figure called a weight relating to the item itself in order that a representative average may be obtained. An extreme example is the average fare paid by passengers on a bus service.

Weighted battery: Refers to a group of educational or psychological tests wherein the relative importance of each test has been determined by attaching a weight to the score obtained in that test.

Weight bias: Bias, usually in an index number, due to the use of incorrect or undesirable weights. . As the true value of the complete quantity which an index purports to measure is not in general capable of direct measurement, bias in this sense has been to some extent an arbitrary quantity.

Weight function: A non-negative function which has been used for weighting purposes especially in the theory of decision functions, where the word has been often used synonymously with 'loss function'.

Weighted index number: Refers to an index number in which the component items have been weighted \according to some system 4 weights reflecting their relative importance. In one sense nearly all index numbers have been weighted by implication; for example, an index number of prices amalgamates prices per unit of quantity and the size of these units may vary from one commodity to another in such a way as to constitute weighting. It has been however, usual to describe an index as weighted' only when weighting coefficients enter explicitly into its definition and calculation.

Weighting coefficient: The coefficient which is attached to an observation as its weight in a procedure involving weighting.

White noise: In analogy with the continuous energy distribution in white light from an incandescent body, a covariance stationary stoachastic process which is having equal power in all frequency intervals over a wide frequency range is termed as white noise. The sequence (X(t), t>O) is said to be a white noise process if it possesses a constant spectral density function.

Wilks' criterion: Refers to a criterion of general use in multivariate analysis for testing hypotheses concerning multivariate normal populations, specially hypotheses of homogenity in means or dispersions. The criterion essentially is dependent on the ratio of the determinants of two matrices of sums of squares and products; the numerator corresponding to a sum with-classes and the denominator a total sum. It occurs m various form.

Winsorised estimation: Refers to a method of estimating the mean of a relatively small sample of observations by using linear systematic statistics and replacing extreme observations by those next in magnitude. This process gets associated with the name of C. P Winsor who gave this approach and it can apply *to* symmetric or non-symmetric censoring and adjustment.

Within-group sum of squares: In analysis of variance the total sum of squares of the observations from the overall mean can be broken down into two components: (1) the between groups sum of squares and (2) the 'within group sum of squares', The latter has been the total of the sums of squares of the observations in each group about the group mean.

Wold's decomposition theorem: Refers to a theorem, which asserts that any univariate non-deterministic stochastic process can decomposed into a one sided moving average process is and a deterministic process.

Working mean: Refers to an alternative term for arbitrary origin.

Wrapped distribution: Refers to the name for a distribution on the real line that gets wrapped around the circle of unit radius. If x has been a random variable on the line with distribution function $F(x)$, the random variable x_w of the corresponding wrapped distribution has been given by

$$x_w = x(mod\ 2\pi.)$$

and the distribution function of x_w has been given by

$$F_w(\theta) = \sum_{k=-\infty}^{\infty} (F(\theta + 2\pi k) - F(2\pi k)$$

Wrapped normal distribution: Refers to the probability distribution resulting from wrapping a Normal distribution on the real line around the unit circle.

'W' Statistic: The test statistic for a rank sum two sample test on dispersions proposed by Ansari & Bradley. It has been generally written as W = ΣR(Z) which has been, effectively, the sum of ranks associated with on sample oi observations after the combined samples get ranked.

X

X-axis: The horizontal axis of a diagram.

Y

Yates' correction: Refers to a continuity correction which may be applied when using the chi-squared test on data from a discrete distribution. It has been generally used only in tests on 2 x 2 contingency tables with small frequencies. It is applied by subtractng 0.5 from the values, of O which are greater than E and adding 0 5 to the values of O which are less than E. (O and E are the observed and Expected frequency respectively in the four categories).

Yates-Grundy Estimator: For a sampling design which has the property that the probabilities of any pair of elements occurring together in the sample is positive, unbiased estimators of the variance of an estimated mean or total have been available.

Y-axis: The vertical axis of a diagram.

Y-intercept: A constant for any given straight line, whose value represents the predicted value of the Y variable when the X variable has a value of 0.

Youden square: An experimental design which was proposed by Youden in 1937. It is not a square, and would be better known as a 'Youden design'.

Yule distribution: Refers to a distribution which was proposed by Yule for biological species investigations of the form

$$f(i) = AB(i, \rho+1)$$

where A and p has been constants and $B(i, \rho+1)$ has been a beta function. Its use was extended by Simon.

Yule's equation A name which is sometimes given to an autoregressive equation of the second order, e.g.,

$$u_1+au_{t-1}+ \beta u_{t-2}=\varepsilon_1.$$

X

X-axis: The horizontal axis of a diagram.

Y

Yates' correction: Refers to a continuity correction which may be applied when using the chi-squared test on data from a discrete distribution. It has been generally used only in tests on 2 x 2 contingency tables with small frequencies. It is applied by subtractng 0.5 from the values, of O which are greater than E and adding 0 5 to the values of O which are less than E. (O and E are the observed and Expected frequency respectively in the four categories).

Yates-Grundy Estimator: For a sampling design which has the property that the probabilities of any pair of elements occurring together in the sample is positive, unbiased estimators of the variance of an estimated mean or total have been available.

Y-axis: The vertical axis of a diagram.

Y-intercept: A constant for any given straight line, whose value represents the predicted value of the Y variable when the X variable has a value of 0.

Youden square: An experimental design which was proposed by Youden in 1937. It is not a square, and would be better known as a 'Youden design'.

Yule distribution: Refers to a distribution which was proposed by Yule for biological species investigations of the form

$$f(i) = AB(i, \rho+1)$$

where A and p has been constants and B(i, ρ+1) has been a beta function. Its use was extended by Simon.

Yule's equation A name which is sometimes given to an autoregressive equation of the second order, e.g.,

$$u_1 + au_{t-1} + \beta u_{t-2} = \varepsilon_1.$$

Z

Z-distribution: Refers to the distribution of a logarithmic transformation of a variance ratio which has been due to R. A. Fisher. If there have been two independent estimates of the Population variance s_1^2 and s_2^2 based upon n_1 and n_2 degrees of freedom the function would be defined as

$$z = \frac{1}{2} \log, \frac{s_1^2}{s_2^2}$$

Fisher selected the transformation of the variance ratio to z to simplify interpclation in tables of significance points.

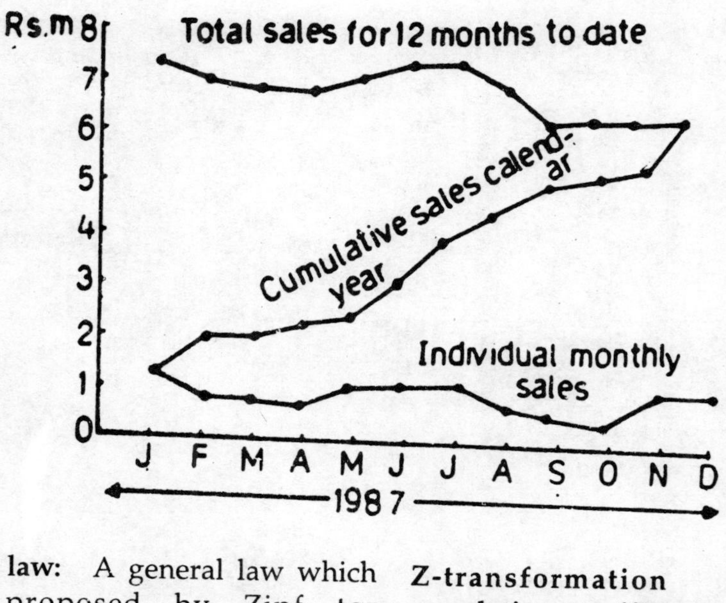

Zipf's law: A general law which was proposed by Zipf tor approximate to the distribution of individuals bearing variate

Z-transformation of the correlation coefficient: A transformation of the correlation coefficient using the formula

$$z = \tanh^{-1} r$$

and taking into account the fact that the distribution of z for samples from a bivariate normal population approaches the normal distribution more quickly than does r.